Edevaldo Aparecido Souza
Jean Carlos Vieira Santos
Lorranne Gomes da Silva

Cultural Geography

Edevaldo Aparecido Souza
Jean Carlos Vieira Santos
Lorranne Gomes da Silva

Cultural Geography

In the urban space of Quirinópolis

ScienciaScripts

Imprint

Cover image: www.ingimage.com

This book is a translation from the original published under ISBN 978-3-330-76742-3.

Publisher:
Sciencia Scripts
is a trademark of
Dodo Books Indian Ocean Ltd. and OmniScriptum S.R.L publishing group

120 High Road, East Finchley, London, N2 9ED, United Kingdom
Str. Armeneasca 28/1, office 1, Chisinau MD-2012, Republic of Moldova, Europe
Managing Directors: Ieva Konstantinova, Victoria Ursu
info@omniscriptum.com

Printed at: see last page
ISBN: 978-620-8-57708-7

SUMMARY

PRESENTATION

This publication is the result of guidelines for Course Conclusion Work (monographs) in the field of Cultural Geography, defended in 2014 and 2015 in the Degree Course in Geography at the State University of Goiás, Quirinópolis Campus.

Quirinópolis is located in a strategic area for agrarian capital, represented by the sugarcane sector, due to the strong natural potential needed for the production process of the two ethanol, sugar and electricity agro-industries. This production process has dynamised the socio-cultural practices of the population of Quirinópolis, who have come to live with new technological demands and other needs of modernity.

However, they bear witness to elements from the past, which represent cultural legacies of this community in the present, with new productive circuits and spatial rearrangements and reinforce other forms of culture and leisure that establish the coexistence and reconfiguration of landscapes and new forms of adequacy and links with place, territory and space. In this way, the aim of this book is to socialise the geographical knowledge acquired through the research of Geography students and teachers, valuing traditional practices, even if they are reinvented, and the knowledge/doings of local people.

Chapter 1 "Knowledge and practice of popular medicine in Quirinópolis: the space of roots at the CEAQUI fair" presents the need to value and preserve traditional medicinal practices. The text presents the research with a subject who knows countless plants, combinations and other traditional healing practices and, based on this experience, the spatial organisation of the fair was analysed, as well as the relationship and reorganisation of the Cerrado space, based on the processes and uses of this biome by the subjects who have settled there for many decades, whose lived and conceived space is related to the production of livelihoods. However, this knowledge is threatened by the advance of agribusiness into the Cerrado, which consequently jeopardises the know-how of traditional populations.

This chapter was written by Ana Paula Ferreira Braga and Edevaldo Aparecido Souza. Ana Paula Ferreira Braga has a degree in Geography from the State University of Goiás, Quirinópolis Campus (2015) and Prof. Edevaldo has a degree in Geography from the Federal University of Mato Grosso do Sul (UFMS), Três Lagoas Campus. Prof Edevaldo has a degree in Geography from the Federal University of Mato Grosso do Sul (UFMS), Três Lagoas Campus (1994), a Postgraduate **Diploma** in Geography, also from UFMS (1999), and a Master's degree in Geography from the São Paulo State University (UNESP) Júlio de Mesquita Filho, Câmpus Presidente Prudente (2005), a PhD from the

Federal University of Uberlândia (UFU), Câmpus Uberlândia (2013) and a Post-Doctorate from the Federal University of Mato Grosso do Sul, Câmpus Três Lagoas. He is a full professor at the State University of Goiás (UEG), Quirinópolis Campus, with exclusive dedication. He has experience in Geography, with an emphasis on Agrarian Geography and Cultural Geography, working mainly on the following themes: Peasantry, socio-cultural practices, transformations in the countryside, Cerrado, sugarcane production, changes in the landscape, territories and ways of life.

Chapter 2 "Handicrafts and craftswomen in urban Quirinópolis: a form of work with meanings that seek means to achieve a certain end" presents and discusses handicrafts as manual labour in the city of Quirinópolis. In this sense, the aim was to survey the handicrafts produced, such as crochet, embroidery, cord and patchwork rugs and gourds with paintings and drawings, the places where this know-how is sold and, above all, the individuals who, with their creativity, reinvent an art with particular and popular traits. The geographical discussion starts with the geographical category of place, moving on to commercial territories.

Marta Correa da Silva has a degree in Geography from the State University of Goiás, Quirinópolis Campus. Professor Jean Carlos Vieira Santos has a post-doctorate in Tourism from the Faculty of Economics at the University of Algarve/Portugal (2014-2015); a doctorate from the Institute of Geography at the Federal University of Uberlândia (2010); a PDEE/Capes internship at the University of Algarve/Portugal (2008/2009); a master's degree from the Institute of Geography at the Federal University of Uberlândia (2004); a specialist in Geography from the São Luiz School of Education/SP (2001) and a degree in Geography from the Federal University of Uberlândia (1994). He is a member of the Association for the Defence of the Cultural and Environmental Heritage of the Algarve in Portugal (ALMARGEM). He is currently a full-time lecturer and researcher at the State University of Goiás - UEG, Caldas Novas Campus (from 2004 to 2016 he was linked to the Quirinópolis Campus). Researcher Incentive Grant (BIP) from the Pro-Rectory for Research and Postgraduate Studies at the State University of Goiás (PrP|UEG). He has published articles in journals and conference proceedings in Brazil and abroad.

He has experience in the following areas: Geography of Tourism; Pottery and Popular Crafts in Tourist Destinations; Leisure and Tourism in Tourist Cities; Territories and Hotel Spaces.

Chapter 3 "Practice, spatialisation and profile of skateboarders in the city of Quirinópolis/GO in 2014 and 2015" shows the spatialisation and profile of skateboarders in the city of Quirinópolis, which is now globalised as a sport and leisure practice that provides health benefits for the body and mind. Four main points and six secondary points for skateboarding in the city were observed. These skateboarders have very strong bonds of friendship with each other, even helping each other out. They act strongly in the urban world, as they are agilising agents of the city-space and belong mainly to the upper-middle, middle and lower-middle social classes

José Ricardo Fonseca Dantas has a degree in Geography from the State University of Goiás, Quirinópolis Campus (2015). Prof Lorranne Gomes da Silva has a PhD in Geography from the Federal University of Goiás (2016), a Master's degree in Geography from the Federal University of Goiás (2010) and a degree in Geography from the State University of Goiás, Cora Coralina Campus (2008). She is currently a lecturer and researcher at the State University of Goiás - UEG, Cora Coralina Campus (until 2016 she was linked to the Quirinópolis Campus). She has worked mainly on the following subjects: Geography Teaching; Didactics and Practice in Geography; Supervised Internship; Scientific Methodology; Cultural Geography; Indigenous Education; Theory of Geographical Knowledge; History of Geographical Thought; Geography of Brazil; Geography of Goiás, Indigenous Peoples.

Chapter 4 "Célio Moraes Andrade Square in Vila Promissão: a place of diversity and contradiction in ways of living" presents the context of the square as a collective space, a synthesis of complex urban life and as a place that comprises a multiplicity of dynamic and complex relationships that are developed between people and the square and between people and other people and reproduced in the sensitive space of what is lived. The text was dedicated to the study of the square in its uses, forms, contents, practices and connections established in the daily life of this place by the actions of social subjects: residents, regulars and market-goers.

Onilda Pereira de Jesus has a degree in Geography from the State University of Goiás, Quirinópolis Campus (2015). Prof Marlane Aparecida Cabral has a degree in Geography from the State University of Goiás (2005).
She is currently a lecturer in Higher Education at the State University of Goiás, Quirinópolis Campus. She has experience in Geography, with an emphasis on Geology, Geomorphology, Hydrogeography and Biogeography.

"Tempero de boteco e a geografia das tradições: no espaço urbano de Quirinópolis (GO)" is the title of the fifth chapter by Silea Xavier Camilo, who has a degree in Geography from Universidade Estadual de Goiás, Câmpus Quirinópolis (2015) and Edevaldo Aparecido Souza.

The text reflects on the history of traditional establishments in Quirinópolis, with the aim of rescuing their history and demonstrating their importance for the urban and cultural development of the municipality, since this type of business existed in times gone by and currently retains the cultural traits of the time. In addition, they contribute to the preservation of a commercial segment, bars/botecos, which maintains its traditional way of life, remaining in the same activity, in a region that has expanded in recent years. It is believed that the secret lies in the seasoning of each typical dish of the places under study, in the relationship with customers, in the simplicity of the place, in the traditions that are passed down from generation to generation.

Analysing the traditional commerce that exists on Rua Rio Preto in the city of Quirinópolis and understanding the social and economic relations that still exist is the aim of the sixth chapter, "Rua Rio Preto in the central area of the city of Quirinópolis: commercial tradition and merchant subjects". This text was written by Renata Dias Borges, who has a degree in Geography from the State University of Goiás, Quirinópolis Campus (2014), in partnership with Jean Carlos Vieira Santos, who investigated the traditional commercial relations of Rua Rio Preto, the subjects involved and the forms and functions of this trade, in view of the provision of new spaces for commercial reproduction in the Quirinópolis intraurban area.

The geographical discussion based on the concept of space allowed us to arrive at the spatial organisation of the people involved with the old commercial enterprises in this part of the city, relating them to the local historical and traditional commercial content, with a view to new approaches and dialogues about the preservation and revitalisation of old urban spaces.

The seventh chapter, "Spatial reorganisations in the municipality of Quirinópolis/GO and female prostitution in 2013", presents a reflection on the spatial reorganisations of the municipality of Quirinópolis and female prostitution, a profession that is so old and ignored, even by the public authorities. The category of space/spatialisation underpins the geographical discussions on the subject, seeking to understand the relationship between the development that the city has undergone with the arrival of agribusiness and the increase in female prostitution.

Lucas Clara Leão has a degree in Geography from the State University of Goiás (2014) and currently works as a communications officer at the State University of Goiás - Quirinópolis Campus (2015). He is the author of this chapter, together with Lorranne Gomes da Silva.

We must express our gratitude to the State University of Goiás (UEG), which made it possible to carry out the research and write the scientific texts. In this way, this publication expresses our gratitude for the institution's investment in teaching, research (and extension), in which the authors of this book are (and have been) involved. We would also like to thank the authors who dedicated part of their time to organising and adapting the content to chapter format as a way of socialising the research and debates throughout the construction of the monographic work.

Organisers

PREFACE

Brazil is indeed a large, diverse and surprising country. In every corner of its immense territory, human life unfolds from the most varied socio-environmental and cultural mediations, generating different, unequal landscapes full of charms and contradictions. But until recently, not all of Brazil's regions and nooks and crannies attracted the interest of Cultural Geography analyses.

In other words, the geography of a non-metropolitan Brazil, not linked to the country's command centres, began to be projected with greater emphasis just a few decades ago, forcing us to come across corners of the country that are veritable blank canvases on which everything has to be written.

In what is known as central Brazil, in the Cerrado domain, lies Quirinópolis, a city in the state of Goiás that has been surprising geography researchers due to its economic dynamism, strongly related to sugar and alcohol production, but also due to its cultural traditions, both urban and rural. It also attracts attention because of the city's rapid pace of technical densification, which is gradually giving the landscape of Quirinopolis an air of post-modernity.

The book we are presenting here represents these contradictions and contrasts well. In it, Quirinópolis is presented as the locus of the study of various professors and students from the State University of Goiás (Quirinópolis Campus), who have endeavoured to present the results of their research into cultural geography. The book is justified by the need for geography to record the most striking cultural features of Brazil's medium-sized municipalities. Much has been written about large urban centres, but there is a huge gap in the literature about small and medium-sized towns.

It is in this sense that this book contributes, as it presents the place of Quirinópolis as it is seen, analysed, described and given new meaning by the people who live there. It is this scientific approach seen "from the inside" that has set the tone for the new Cultural Geography movements in Brazil. "Being from" (or "living in") Quirinópolis and writing about this municipality provides a very special context for the articles in this book, as its authors can argue more appropriately, with a sense of belonging and effective connection. It's gratifying to see the geography of places being done by the geographers of those same places.

In their work, the writers took great pains to deal with themes such as regional folk medicine, handicrafts and other subjects that are classic in cultural studies. But the city's streets and squares also came up in the research, as places for urban sports, prostitution, the market and life in its broadest senses. In this research scenario, the categories of **territory, landscape** and **place** were the most visited by the geographers who studied Quirinópolis, showing nuances of the city that represent true memorial records, given the novelty of the empirical approaches.

For Quirinópolis (in the **strict** sense) and for Goiás (in the **broad** sense), the work's contribution is very significant. UEG is fulfilling its role, giving the people of Quirinópolis the return they expect from an institution committed to the community in which it operates. Goiás geography gains a new and innovative work and Brazil has the opportunity to better understand its inland municipalities, which are not always the target of high-impact cultural research.

The authors, teachers and students are to be congratulated. The organisers are also worthy of merit and recognition, as we know that undertaking projects of this nature is no easy task, especially in difficult times like the present.

Finally, we hope you enjoy the book we have just presented. The texts are excellent, well-founded, well-illustrated, with important theoretical contributions and case studies. Some are richly illustrated with good quality charts, graphs and maps that will help readers understand the topics covered.

May we soon be able to count on new contributions from the geographers of Quirinópolis, with other publications that provide us with knowledge and disturbing themes of the regional reality.

Prof Dr Anderson Pereira Portuguez
Postgraduate Programme in Geography at Pontal
FACIP - Federal University of Uberlândia.

CHAPTER 1

KNOWLEDGE AND DOINGS OF POPULAR MEDICINE IN QUIRINÓPOLIS: the space of roots at the CEAQUI fair

BRAGA, Ana Paula Ferreira [1]
SOUZA, Edevaldo Aparecido[2]

Introduction

This text is part of the monograph presented and defended in 2015 in the Geography Degree Course at the State University of Goiás (UEG), Quirinópolis Campus. Research into medicinal herbs is an inconclusive endeavour, as discoveries appear over the years, innovating healing methodologies. The aim of this research was to get to know the daily life and know-how of Mr Lázaro Francisco Dias, one of the few raizeiros still in existence in the city of Quirinópolis, to whom hundreds of clients who, combined with medicinal herbs and faith, turn for a cure for their ailments. In the past, this practice of healing with plants was more popularised, but today it is no longer so common, although it is known that it is a less aggressive medicine.

Lázaro Francisco Dias was born on 27 June 1956 in the municipality of Caçu-GO; he came to the municipality of Quirinópolis at the age of ten, where he was raised; he then moved to some towns in the region, such as Jataí and Paraúna; he returned again to Quirinópolis and, years later, in 2006, he moved to Castelândia, where he lives. He is currently an evangelical of the Congregação Cristã do Brasil church.

Mr Lázaro sells his products in Quirinópolis and neighbouring municipalities. Roots, leaves, bark and fruit are collected in the Cerrado region. Castelândia is in the south of the south-western micro-region of Goiás and Quirinópolis is in the micro-region of the same name (Figure 1).

[1] Graduated in Geography from the State University of Goiás, Quirinópolis Campus. paulinhabraga@hotmail.com.
[2] Professor of Geography at the State University of Goiás, Quirinópolis Campus. .ediueg@gmail.com

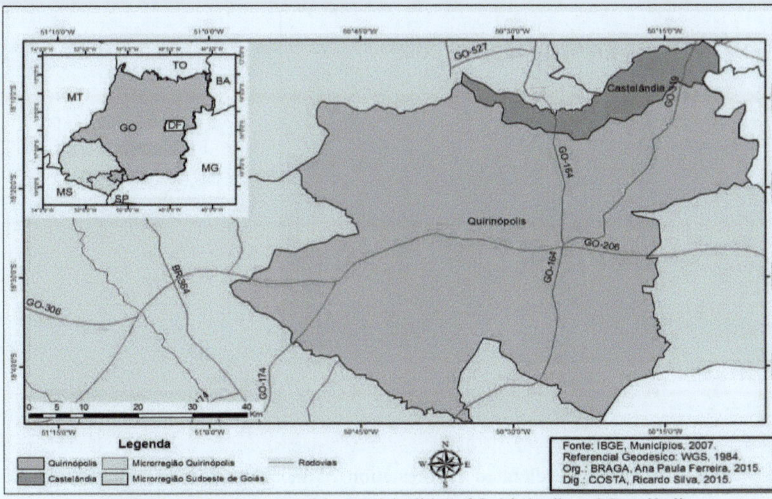

FIGURE 1: Location map of the municipalities of Quirinópolis and Castelândia
Source: IBGE, Municipalities (2007). Org.: BRAGA, A. P. F (2015); Dig.: COSTA, R. S. (2015)

Even though he lives in Castelândia, this root grower found the Quirinópolis Central Supply and Leisure Centre (CEAQUI), better known as the Covered Market (Figure 2), to be the right place to spread his products and his medicinal herb know-how. Since 1998, he has had a home remedies stall at the fair on Sundays.

CEAQUI has been growing, both in terms of the market traders and the market-goers who, with their families, shop for Sunday and weekday lunches. Many still go to the market for a simple stroll or to meet up with friends. According to Pereira, Garcia and Oliveira (2009, p. 12):

> The Central de Abastecimento e Lazer de Quirinópolis - CEAQUI (Quirinópolis Supply and Leisure **Centre),** located at Av. José Vicente de Paula s/n° - **Centro, popularly known as the "Quirinópolis covered market", sells** fruit and vegetable **products on Sundays,** mainly from local and regional family production. During the week, some of the traders in these products, known as "feirantes", trade permanently at other points throughout the city of Quirinópolis.

This market receives producers from the city and neighbouring regions, supplying the residents of Quirinópolis and the surrounding area - as well as being a space for social, economic and cultural relations. From this premise, the research sought to show that the use and knowledge of medicinal plants is still present today, from those who have the science of processing the herbs, such as Mr Lázaro, but also from the population who seek them out, since he inherited the custom of using these medicines from the popular wisdom .

FIGURE 2: CEAQUI - Quirinópolis Free Trade Fair
Source: BRAGA, A. P. F. (2015)

Even with the accelerated deforestation of the Cerrado, we still have these practices among the population. However, we need to be concerned about the extinction of this biome, as medicinal plants and the knowledge of how to process medicines will also disappear as a result.

In order to achieve the proposed objectives, the research began with a discussion of the geographical category of **space**, given that the entire work process of the subject under investigation has its own peculiar way of organising space. The survey of theoretical references based on published books and articles, fieldwork and photographic surveys were also fundamental to the construction of the work.

The Feira Livre as a space for preserving medicinal culture

In the quest for human survival, food stands out, but the extraction of herbs for folk medicine and home remedies from teas and bottles has also always been important. There are various ways of producing and commercialising homemade remedies, depending on the way of life or other forms of culture and belief of each individual or social group, in other words, there are various popular traditions that are established in people's daily lives and that contribute a collective cultural development. According to Horta, Grunberg and Monteiro (1999, p. 7), there are many forms of cultural expression:

> constitute the living heritage of Brazilian society: handicrafts, ways of fishing, hunting, planting, cultivating and harvesting**, of using plants for food and medicine,** of building houses, cooking, dances and music, ways of dressing and speaking, religious and popular rituals and festivals, social and family

relationships, reveal the multiple aspects that a community's living and present culture can take on. (emphasis added).

Cultural expressions are set in a certain space, which is linked to man and nature, which are subdivided into the social and cultural environment where man looks for sources of survival that relate to people's ways of life in general. According to Moraes and Costa (1984), the relationship between man and nature occupies a prominent place in the understanding of work.

Space is organised according to man's relationships with nature and the countless resources that can be taken from it to produce important elements in the social organisation of a given population, such as home remedies from many plants and roots. According to Moraes and Costa (1984, p. 131), the "society-space relationship is understood as a process of valorisation, its internal movement must express, in essence, a permanent process of creating values. However, this is distinct from the valorisation of capital".

Space is organised according to the relationships between man, nature and the countless resources that can be extracted from it to produce important elements in the social organisation of a given population, such as home-made remedies from many plants and roots. Man, in fact, places a means of organisation, commercialisation and consumption in this space and the way he deals with these remedies also expresses the faith that each person places in this artisanal way of healing.

The valorisation of space as a mode of production and, consequently, a remodeller of society in a process of human labour, can be reconfigured with each cultural tradition. Moraes and Costa (1984, p. 123) state that "[...] the society-space relationship is, from the outset, a value-space relationship, since it is substantiated by human labour".

According to Santos (1988, p. 64), the production of space is the result of men acting on space itself through natural and artificial objects. "Space is not, therefore, as in the classic definitions of geography, the result of an interaction between man and raw nature, nor even an amalgam formed by today's society and the environment". The relationship between nature and society is

mediatised by work. For this author, "space must be considered as an inseparable whole in which, on the one hand, a certain arrangement of geographical objects, natural objects and social objects participate, and, on the other, the life that fills them and animates them, in other words, society in movement". (SANTOS, 1988, p. 26).

The root grower at the Quirinópolis Free Fair, Lázaro Francisco Dias[3] , has a very peculiar, deeply elaborate spatial organisation that serves the purposes of producing the medicines and roots he sells. The way in which the space is organised and occupied is carried out in a very traditional way by the root grower, who makes the Castelândia-Quirinópolis journey from 4am, i.e. in the early hours of the morning, Brasília time. He arrives at the market at around 5.20am, sets up his stall and begins his day's work. He collects from friends' farms in Castelândia and Quirinópolis and often from the side of the road.

From this context, nature is also included in this organisation of man-space. In fact, it provides the necessary elements for production, such as food, clothing and human health. Nature is therefore seen as the source of the natural resources used by the subjects to make these home remedies, which are vitally important for culture and tradition. It is necessary to emphasise that man is part of it, yet he is a being who transforms it in order to satisfy his basic survival needs. According to Borges (2011, p. 130),

> Throughout history, nature has been an element of strategic importance due, among other factors, to its potential to provide raw materials, especially for food, agriculture and **human health**. In contemporary times, its relevance must be considered, especially for **traditional populations** who identify a particular way of maintaining their natural resource base, with its historical and cultural heritage, in other words, an ecological culture. (emphasis added).

Nature, for Mr Lazarus, means the structure that holds the earth together. "We are moved by nature, if it comes to an end, we go first. Everything that goes on our table comes from nature. God maintains nature". He adds that the raizeiros and all living people depend on nature. "Harvesting and working with herbs requires a lot of experience, which you only acquire as you work. I've been working in the area for 18 years now, and it's been 18 years of teaching. The greatest teacher is nature itself" (verbal information).

Fundamentally, nature becomes relevant to the production of these remedies and the organisation and use of space by the subjects in an artisanal way. Both harvesting and manufacturing make it possible to produce products and also to sustain themselves . According to Dollfus (1982, p. 48):

> When analysing the relationship between man and his environment, it is essential to study the very complex role played by the environment created and produced by societies on those societies themselves and on the individuals who

[3] Mr Lázaro, the protagonist of this research, agrees and makes a point of having his name revealed in this text, because it records the memory and history of his life and activity with medicinal herbs. The authorisation to disclose his name and images is duly documented in a Consent Form, stored in our archives.

make them up. The environment in which man lives in countless communities is becoming less and less natural. Geography does not neglect the study of these interactions between man and his work.

In this sense, the Cerrado is nature for Mr Lázaro and it is necessary to preserve this biome so that the knowledge and extraction of herbs can continue with this and other raizeiros. The Cerrado is the support for the maintenance of so many traditional peoples' know-how. Natural remedies are just one of them.

The Cerrado as the support and maintenance of medicinal traditions

The Cerrado is important to the population, which depends on it for its relationships, especially when it comes to gathering food and medicines. You can't talk about home remedies without relating them to the study of the Cerrado Biome, both in terms of natural forms and anthropogenic relations, still with the concept that man is part of this nature.

The Cerrado is located on Brazil's Central Plateau and is the country's second largest biome, surpassed only by the Amazon Rainforest. It covers the states of Goiás, Tocantins, the Federal District, part of the states of Bahia, Ceará, Maranhão, Mato Grosso, Mato Grosso do Sul, Minas Gerais, Piauí, Rondônia and São Paulo. It also occurs in separate areas to the north in the states of Amapá, Amazonas, Pará and Roraima, and to the south on small islands in Paraná (Figure 3). The vegetation of the Cerrado biome is characterised by forest, savannah and grassland formations.

According to Souza (2013), the Cerrado biome contains various ecosystem formations classified as: clean fields, dirty fields, cerrado **stricto sensu**, cerradão, gallery forests, veredas and cerrado rupestre.

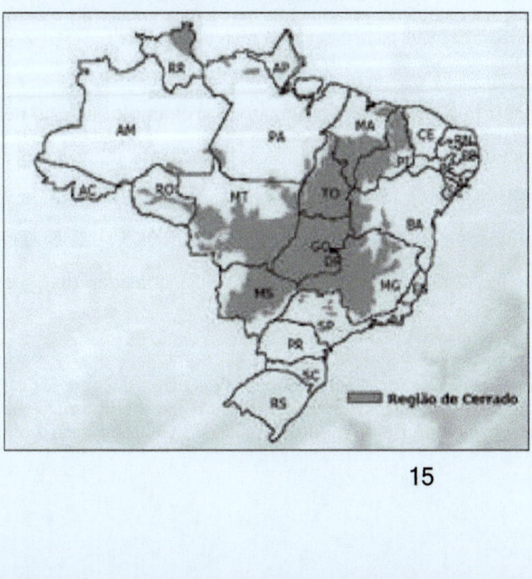

FIGURE 3: Map showing the location of the original Cerrado area
Source: AIBA (2015)

This biome, according to Chaveiro and Castilho (2015, p.2), is considered a heritage that integrates the "life in which the classes of vegetation, the hydrographic basins, the relief, the soil, its space, its culture, its symbols, its people, its art, the different ways of life that have been constituted here participate" (emphasis added). The authors believe that this "is a living and dynamic archive of colours, flavours, sounds, thicknesses, corners and reliefs".

Cândido (1979, p. 23) expresses these relationships well, saying that "the existence of every social group presupposes the achievement of a relative balance between its needs and the resources of the physical environment, requiring, on the part of the group, more or less adequate and complete solutions [...]". It is from these relationships and tensions that the ways of life of a given population are formed.

One of the important characteristics of the Cerrado is its inhomogeneity and biodiversity, made up of various subsystems, forming a natural mosaic with diverse fauna and flora. In this sense, Ferreira (2015, p.7), anchored in Souza's Dictionary of Terminology (1973), defines the Cerrado vegetation as consisting of "spaced, twisted, low trees with crooked branches and thick, ruddy or cracked bark. Between the trees, forming the background, there is a more or less dense stand of grasses and field plants" (FERREIRA, 2015, p.7).

In this way, there are many possibilities and abilities available to individuals and social groups to relate to and draw from these ecosystems the resources they need for their daily activities.

> The socio-cultural elements provide a sense of belonging to the territory, with social ties, the result of the re-signification of meanings, knowledge and ways of using the space itself. Even peasants who have left the community and are living in the city carry with them the cultural traits of that place [...] (SOUZA, 2013, p.186).

In this sense, there are many raizeiros, like Mr Lázaro, for example, who is no longer in the countryside but depends on it to carry out his activities. He inherited the knowledge and skills for making home-made medicines from peasant traditions. "They are like an inheritance of the knowledge and skills kept by the families," adds Souza (2013, p.253). For this reason, Chaveiro and Castilho (2015, p.2) consider that the Cerrado:

> It needs national defence because it is the nation's heritage. It also needs the committed action of educators, researchers, political and social institutions. And it needs to be understood in all its variation, multiplicity and from

> different angles. An integrated analysis of this important living environment [...].

Due to the destruction of the environment, both medicinal herbs and vegetation have been rapidly devastated. As a result, this practice is gradually disappearing, as is folk medicine. According to this author:

> The accelerated destruction of ecosystems and their biodiversity has jeopardised the **survival of medicinal species** used by rural populations, drastically reducing their availability and even extinguishing them completely (CALIXTO; RIBEIRO, 2004, p.2. Emphasis added).

Therefore, deforestation, as Mr Lázaro understands it, is very damaging to nature and to the know-how of the people of the Cerrado, who look to this environment for the raw materials they need. Souza (2014, p.174) corroborates that "it's not easy, the herbs are becoming extinct due to the devastation of the Cerrado". It is this concern with the devastation of the Cerrado that the following subchapter will address.

As well as being the second largest biome in terms of extension, the Cerrado also has a rich fauna and flora, providing a specific biodiversity of the ecosystems formed within this biome. However, deforestation (Figure 4) has generated substantial concern about the extinction of thousands of species, as well as changing the way man relates to nature.

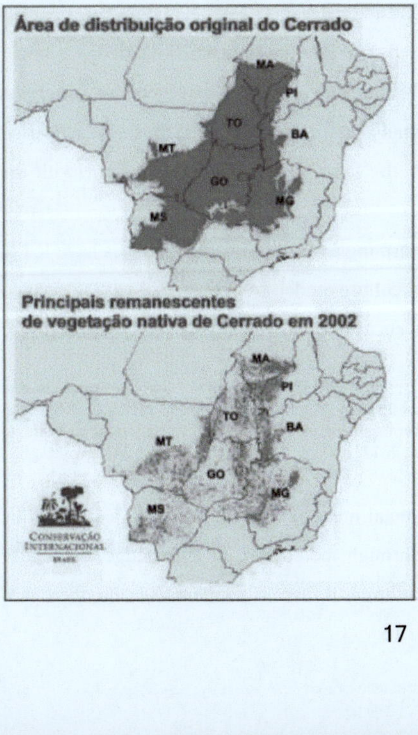

Figure 4: Deforestation of the Cerrado up to 2002
Source: Deforestation in Brazil (2015)

According to Casseti (1995, p.25), these changes are imposed by the capitalist production model, above all by agriculture and agro-industries processed in rural areas, but also by industries installed in urban areas.

> The worsening of environmental problems is therefore born out of private property relations and class antagonisms, which are responsible for changing the root of the social structure and, consequently, the relationship between man and nature. In short, environmental impacts have worsened as a result of the greater anarchic development of the productive forces that structure the capitalist mode of production, while relations of production are relations of domination and submission.

Thus, with the increase in population and technologies, the devastation of these Cerrado areas tends to evolve and leads us to worry about deforestation and the loss of knowledge of these subjects, because it is in nature that man seeks a cure for diseases in general and innovates his know-how, despite the advance of technologies. As Martins (2006, p.9) points out:

> Man seeks **healing in nature, in plants,** for example, but nature itself limits anthropic action in terms of its devastation or degradation. This means that today we have in our hands advanced techniques for discovering cures for various diseases. However, this nature is **threatened** by the intense degradation to which it has been subjected for a long time (emphasis added).

In Souza's (2013) opinion, this is knowledge that has been stored in people's memories to be used at the right time. Faced with this situation, this knowledge is under threat. The vegetation where man harvests his plants for traditional medicine; where he takes food supplements such as fruit, chestnuts, guarirobas and others; and where he used to grow dozens of agricultural plants for the family consumption economy (with the surplus being sold), is being threatened.

Since the 1960s and 70s, modern farming and cattle-raising practices have led to these activities being replaced by the monoculture model, from the people of the cerrado to the "seas" of soya, corn, sugar cane and eucalyptus. According to Souza (2013, p.77),

> The Brazilian government's development policies have resulted in the deforestation of this biome's vegetation, a reduction in the number of small properties and the displacement of peasant families and rural workers to monoculture (SOUZA, 2013, p.77).

In fact, nature is the most fundamental means by which man can perpetuate his traditional knowledge of folk medicine through the sharing of knowledge, especially through the teachings of parents to their children. If there is no more Cerrado, there will be no reason to pass on this knowledge and so it will be lost.

For Borges and Almeida, "the raizeiros have an important role to play in potentiating their popular knowledge of medicinal plants from the Cerrado, in a systematic activity of producing medicines and remedies" (BORGES; ALMEIDA, 2012, p.2).

In this way, people are forced to maintain the Cerrado in order to find and maintain their knowledge so that they can make medicines. However, even these extractive practices can be harmful to the environment, especially when it comes to collecting bark which, if not properly managed, can kill the tree.

In this way, Mr Lázaro collects his roots and leaves in an indeterminate quantity, depending on sales and demand for the herbs. The collection period is during the dry season and the waning moon, because during the rainy season the herb takes longer to dry and during the waning moon the plant's sap is at its best. The roots are bagged in plastic bags on the shelves at the top. Some are

difficult to access. The root of the amaroleite, for example, is found in the mountains and you have to dig approximately 1.30 metres deep to remove the potato.

There are various ways of impacting nature and medicinal uses, not just deforestation, but also erosion, water contamination, desertification, soil impoverishment, among others. According to Barros (2009, p.10):

> The damage produced by such human actions is not only felt by nature, but society in general ends up suffering because of the attitudes of some of its members. For nature, some consequences are more costly: the impoverishment of the soil, resulting from the occupation of agriculture through the release of fertilisers and the planting of single crops, ends up making the soil incapable of recovering its nutrients, causing desertification, for example. Erosion is another major threat posed by human occupation, since soils are exposed to environmental factors that can cause erosion, and often, in addition to damaging nature, end up directly affecting social life there.

For Souza (2013), the relationship between social groups and the Cerrado is based on the various elements it has to offer. However, for this study, vegetation is the most important, as it is from it that Mr Lázaro obtains and produces the wide variety of medicines he sells. In this sense, according to Souza, this vegetation,

> As well as being the predominant element in the composition of the landscape, it also mediates the rural population's contact with fauna. It is part of the food system and, above all, part of the **popular medicinal wisdom of the Cerradeira**. Thus, **deforestation and the establishment of a monoculture plantation mean**, above all, for this population, the **loss of their medicines**, animals and complementary food (SOUZA, 2013, p.140. Emphasis added).

Deforestation for agricultural production and export has resulted in considerable losses of the Cerrado's native vegetation and its biodiversity, as well as its popular medicinal maintenance. According to Souza (2013, p.255),

> As **there is almost no Cerrado vegetation left, traditional knowledge tends to disappear,** as the scarcity of fruits, which complemented food knowledge, and the **varied herbs that nourished the medicinal practices of Cerrado peasants**, is evident with the large-scale expansion of agribusiness. The near elimination of anjico *(Anadentherafalcata),* jaracatiá *(Jacaratiaspinosa),* gabiroba or guaivira *(Campomanesiacambessedeana)* and jatobá *(Hymenaeacourbaril),* goiabinha do Cerrado *(Psidiumfirmum Berg)* cajuzinho do Cerrado *(Anacardiumothonianum),* mama cadela *(Brosimum Gaudichauii),* marmelo *(Cydonia oblonga Mill),* chapéu de couro *(Echinodorusgrandiflorus)* sucupira *(Pterodonemarginatus).* quina *(Cinchona calysaia)* and other Cerrado trees and shrubs greatly jeopardise the know-how and studies of this knowledge of the place, in the place and for the place (emphasis added).

Chaveiro and Castilho (2015) show that in Minas Gerais, where there are many coal-fuelled industries, there has been and still is intense deforestation for coal production and that in the states of "Goiás, Mato Grosso do Sul and Mato Grosso, in addition to coal production, the removal of vegetation makes plantations possible". These activities, together with livestock farming, take place in areas of intense activity linked to the agribusiness sector.

According to Souza (2013), with the reduction of the Cerrado's biodiversity, and as a consequence, in addition to the difficulty or even impossibility of using medicinal knowledge, no matter how simple the illness, the population has become dependent on chemical medicines prescribed by doctors, professionals who are difficult to reach for consultations. For Souza, "there needs to be a 'vitalisation' of a historical know-how that has accompanied them since ancient times and which must not be forgotten, otherwise it will disappear" (SOUZA, 2013, p.261).

Despite this, Mr Lázaro understands that this practice of deforestation constitutes a threat, because in the future this culture of folk medicine could end, and he may not be able to cope with this loss of knowledge, since this change in the ecosystem is causing the medicinal herbs to disappear.

Therefore, the know-how of the raizeiro and the people of the Cerrado is at the mercy of the devastated environment, the machines and the agricultural processes that man generates in the environment, making it difficult to harvest these medicinal herbs. However, with the intervention of practices such as those of Mr Lázaro, it is possible not to let folk medicine fall into oblivion.

The CEAQUI Free Fair in Quirinópolis, therefore, is a fundamental space for the preservation of popular memory and for the simplest and cheapest ways of combating human diseases and even those of some animals.

The Feira Livre in Quirinópolis is of great importance to market-goers. It is a great support for peasant family farmers, as they sell their small products and buy them from other producers at the same time. In other words, a practice with the characteristics of short agri-food chains[4] .

For consumers, the open-air market is an excellent opportunity, because it's where they can find fresh, natural products, picked from the gardens, as well as others such as sweets, fruit and vegetables, especially for the subject of this research, Mr Lázaro. His medicine stall contains roots, leaves and other substrates from the Cerrado vegetation.

For Quirinópolis, the preservation of the market is of great importance, since on the one hand it generates income for the market traders, and on the other hand it provides Quirinopolinos with the opportunity to consume fresher food, as well as fostering relationships for meetings, outings and other human activities. The market is very popular with people from the city and neighbouring regions, and for many it's a family atmosphere. According to Pereira, Garcia and Oliveira (2009, p. 25):

> The market itself attracts a wide range of consumers and is visited by people from various neighbourhoods and the surrounding regions, who have traditionally come here for a long time out of tradition, leisure or habit. This environment is an attraction for those who live here or visit Quirinópolis. It's a place to go for a walk and a means of support for many families [...].

According to information from regulars, before the Feira Livre, the space was used for a dirt football pitch, where games were played between teams and there were even women's teams, but these championships were just for fun at weekends. According to Pereira, Garcia and Oliveira (2009, p. 29), the Quirinópolis Fair emerged with its first market-goers in 1988, occupying the football ground:

> On 8 February 1988, by means of Law 1.514, the covered market was renamed the Quirinópolis Supply and Leisure Centre - CEAQUI. There is no specific date when the market was inaugurated. However, according to unofficial information, it took place thirty days later.

Each stallholder has their own space and their own way of organising them: vegetables, pastries, clothing stalls, among others. Mr Lázaro's stall, with its peculiar way of organising the herbs, is next to the other stalls selling other products, such as food and poultry, clothes and handicrafts (Figure 5).

---

[4] Forms of commercialisation that express proximity between producers and consumers, not only and necessarily in terms of space, but a kind of connection that allows for interactivity, making it easier for both to know each other's purposes (SCARABELOT; SCHNEIDER, 2012, p.103).

The market is synonymous with tradition in the city. The act of going to the market to eat pastel, savour garapa and drink coffee with cheese bread is part of the Sunday routine for people who go out for a stroll, see friends and buy products from the market traders. The culture of going to the market is passed down through the generations, as parents take their children and grandchildren, and some children accompany their parents into old age.

FIGURE 5: Mr Lázaro's stall. In the background, a vegetable stall and a clothes stall
Source: BRAGA, A. P. F. (2015)

Knowing that the market is a place for medicinal herbs, another daily practice of market-goers, also on Sundays, is the search for medicines for a multitude of ailments that afflict families in the city, but also in rural areas.

Know-how and the relationship with the Cerrado

Mr Lázaro is a great connoisseur of the medicinal plants he collects in the Cerrado and knows all about their applications. However, one of his biggest concerns is the deforestation of the Cerrado due to the expansion of agribusiness in the south of Goiás. For him, it's a huge loss because of the damage it does to both nature and human health. There are thousands of diseases that can be cured by medicinal herbs, provided they are used correctly. "This makes me very upset and very worried about deforestation and the extinction of these medicinal herbs."

The reorganisation of Cerrado space for large-scale agricultural production dismantles the entire organisation and processes of Cerrado use by the peoples who have settled there for many decades. It is a lived and conceived space whose specificity is the production of the means of life. It is the replacement of the space of the Cerrado and, consequently, of know-how with flora species, with the space of agribusiness. According to Borges (2013, p. 27), "nature, throughout history, has been an element of strategic importance due, among other factors, to its potential to provide raw materials, especially for food, agriculture and human health".

Knowledge, inherited from times gone by, makes a major contribution to the organisation of space. These same traditions, however, are shaken by the movement of society, but it is still possible to recreate and reinvent what is traditional. The permanence of traditional practices is possible because people are able to recreate the social relationships that are important for reaffirming their ways of life.

There is a sense of belonging that people feel towards their living space. For this reason, Moraes and Costa (1984) state that human thought suggests the valorisation of space, both living and production space. Changes can be more superficial or more intense, as can the relationship between society and space.

The Cerrado is home to a wide variety of plants that, alone or together with others, with just water or other substances, are capable of easing people's suffering. The subject of this research knows hundreds of them well. Some of them will be described below:

The Caroba (Jacaranda caroba) is a tree that grows 8 to 12 metres tall and 20 to 30 cm in diameter. It has small leaves, lilac flowers and capsules that produce "flying" type seeds. It is a blood purifying plant that can be used as an anti-inflammatory and a very mild antibiotic, generally used for infection, and the bark and leaves of the plant are used.

Swallowwort (Euphorbia hirta L.) is a herbaceous plant with a lot of branching, which can vary from thirty to eighty centimetres, with a purple colour (Figure 6). It can be used to treat the heart, unclog veins and also inflammation of the eyes, such as conjunctivitis. All the elements of the plant are used. The root grower even demonstrates that when you break the stalk of the plant, a white liquid oozes out, which is placed directly on the eye that has the disease, in order to improve the infection. The curious thing is that, in the past, it was said that if this liquid got into the eyes, it would cause blindness.

He recounts something that happened when he was twelve years old. His father, working in the fields, stepped barefoot into a blaze and got a third-degree burn on one of his feet, and the burn opened up. A year later, the wound was opening up even more. His father said that if he found an Amaroleite potato (Operculina hamiltonii), as shown in Figure 7, he was sure he could be cured of the wound. His father's brother had shown him where the plant was, a vine with a potato on it. So he went to get the plant. At just twelve years old, it was his first experience of coming into contact with a medicinal plant.

He remembers that his father cut the grass into thirty slices, about one centimetre in diameter and five and a half centimetres thick. He tied them up in a string and put them out to dry, first in the sun and then on the hob of the wood-burning cooker. The slices of grass kept getting that smoke. Every day, he took a slice and put it in a pot to boil with a litre of water. Once the liquid was cold, he drank it throughout the day. He did this for a month and the wound was gone after a year of being open.

FIGURE 6: Swallow weed (Euphorbia hirta)
Source: BRAGA, A. P. F. (2014)
FIGURE 7: Amaroleite (Operculina hamiltonii)
Source: BRAGA, A. P. F. (2015)

The experience has stayed with him ever since. That's why Amaroleite is one of the herbs he still works with today. The problem is that, according to the interviewee, "here in the region it's very difficult [to find it] because of deforestation. It's already very difficult, we still find it on top of a mountain and there's not much of it, it's already very scarce. However, he emphasises that it is a plant that, with a lot of care, can be reproduced.

Algodãozinho do Cerrado (Cochlospermum regium), another herb widely used by people, is also a shrub native to the Cerrado region, like the others already shown, with yellow flowers (Figure 8). The whole root is used as an anti-inflammatory and can be used to treat infections of the uterus, ovaries and kidneys. Barbatimão (Stryphnodendron adstringens), as well as having great healing power, can also be used as an analgesic to reduce the sensation of pain and to treat itching, skin infections and wounds. The bark of the tree is used (Figure 9).

FIGURE 8: Cerrado cotton (*Cochlospermum regium*)
Source: BRAGA, A.P.F. (2015)
FIGURE 9: Barbatimão (*Stryphnodendronadstringens*)
Source: BRAGA, A.P.F. (2015)

Partridgefoot (**Croton antisyphiliticus**) also has great healing powers and is used for everything. It treats infections of the uterus, ovaries, kidneys, discharge and is most suitable for women. For men, it can be used for prostate inflammation. The root, says Mr Lázaro, has to be sold already packaged and certain care has to be taken, as it has great healing power and is very expensive due to the extinction of the species (Figure 10).

In addition to the Cerrado herbs mentioned above, Mr Lázaro also works with dozens of others, such as canfor (**Croton** sp.), castor bean (**Ricinus communis**) and Cerrado liquorice (**Periandra mediterrânea**). The relationships between the subjects that permeate the history and memory built on the knowledge and habit of consuming natural medicines preserve and perpetuate socio-cultural practices and the social reproduction of populations and communities. The indications for use for each plant are shown in Chart 1:

The raizeiro also works with countless other roots to formulate other medicines, such as syrups and garrafadas. The bottles, ready for sale and consumption, are indicated for men as a stimulant and for the prostate, while for women they are usually effective for kidney and bladder problems (Figure 11).

TABLE 1: Family, species, common name, origin and use of some of the plants that Mr Lázaro sells at the market

Family/Species	Common name	Origin	Medicinal use
BIGNONIACEAE			
Jacaranda caroba (Vell.) DC	Caroba	Native	Blood purifying, anti-inflammatory and antibiotic.
BIXACEAE			
Cochlospermumregiu m (Schrank) Pilg.	Cerrado cotton	Native	Infection of the uterus, ovaries, kidneys, more suitable for women.
CONVOLVULACEAE			
Operculinahamiltonii (G.Don) D.F. Austin & Staples	Amaroleite, purga-de-amarelo-leite,	Native	Blood purification, elimination of old wounds on the skin, among other various infections.

	batatão.		
EUPHORBIACEAE			
Crotonantisyphiliticus Mart.	Birdsfoot trefoil	Native	Infection of the uterus, ovaries, kidneys, discharge. Suitable for women. For men, it can be used for inflammation of the prostate.
Croton sp.	Singer	Native	Anti-inflammatory, analgesic and anaesthetic, can be used on insect bites.
Euphorbia hirta L.	Swallow weed	Native	Used to treat the heart, clogged veins and inflammation of the eyes, such as conjunctivitis.
Ricinuscommunis L.	Castor beans	Cultivated	The stone produces an oil and can be used for hair loss, skin wounds, as it helps with healing, and it used to be used as a laxative, but this practice is no longer used.
FABACEAE			
Periandra mediterrânea (Vell.) Taub.	Shredded liquorice	Native	A syrup for bronchitis, pneumonia, coughs and throat infections.
Stryphnodendronadstringens (Mart.) Coville	Barbatimão	Native	It has great healing and analgesic properties, treating itching, skin infections and wounds.

Source: Plants identified by Dr Isa Lucia de Moraes Resende (UEG).

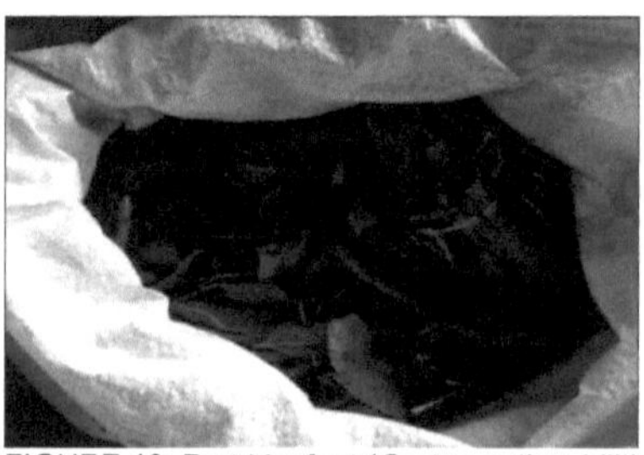

FIGURE 10: Partridgefoot (*Croton antisyphiliticus*)
Source: BRAGA, A. P. F. (2015)
FIGURE 11: Bottles ready for consumption
Source: BRAGA, A. P. F. (2015)

The 16-herb syrup (Figure 12), made from 100% natural ingredients, is recommended for bronchitis, pneumonia, flu, coughs, the respiratory system and pulmonary emphysema.

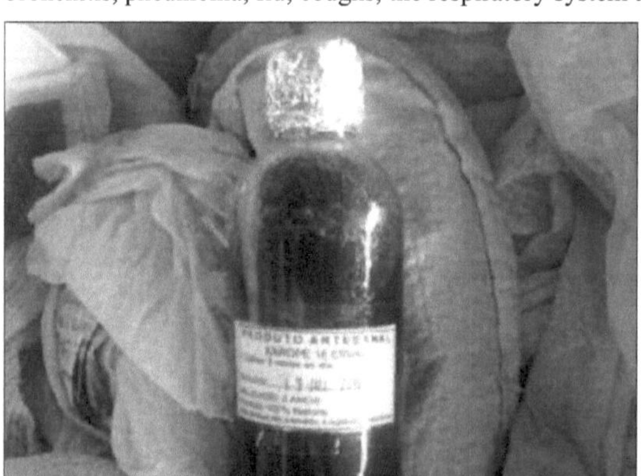

FIGURE 12: Syrup 16 herbs

Source: BRAGA, A. P. F. (2015)

Roots, wood, bark, stems, vines, etc. are all natural health resources found in the Cerrado that can be considered simple plants with great healing powers for the population, but which could become extinct in the very near future, threatened by deforestation.

The subject between history and memory

Mr Lázaro usually gets up at 3.20am, i.e. dawn, and leaves Castelândia in the direction of Quirinópolis, arriving at around 5.20am. He doesn't start setting up the stall until daybreak. Firstly, he erects the mechanical part of the stall, with parts in the form of railings, then the top parts, which form the base for displaying the products, at around 7.30am. Usually, at 12 noon, he starts to dismantle the stall (Figure 13) and, one by one, he puts the same pieces together and ties them with a rubber band. Everything is very well organised and stored in its proper place in the car. At 1pm, she returns home. With the exception of the time spent travelling, he is on his feet from the moment he gets up until the moment he leaves.

In Castelândia, two years ago, he worked selling pequi, an activity he no longer carries out, dedicating himself to working as a market vendor in Quirinópolis. When the day comes to harvest the roots and leaves, the work is doubled: a day in the field and a week at home to prepare the roots in the right conditions for use. Sometimes the stock isn't so low and the preparation time is shorter.

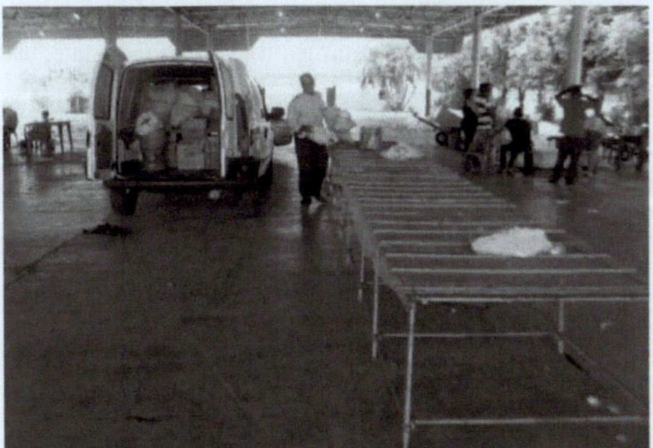

FIGURE 13 - Mr Lázaro dismantling the stall
Source: SOUZA. E. A. (2014)

Therefore, a specific relationship was formed with the space through which he travels: from his home to the market or to the Cerrado and also the organisation of the space in his home for the preparation and handling of medicines. As Braga (2007, p.67) describes, the organisation of space is understood "as a communion between social interaction".

According to the interviewee, his interest in roots and medicinal herbs began when he was twelve years old, but at that age he also contracted very severe anaemia, a disease that, if left untreated, could have led to his death. The anaemia, a disease caused by worms, meant that he was only interested in eating things like roof tiles, charcoal and wood bark. At the time, he lived on a farm and right opposite his house was his grandmother's house, separated from his only by the road.

His grandmother was the daughter of someone who had a lot of knowledge about medicinal herbs, so it was a cultural heritage. Because of this anaemia, his grandmother called him to go and get the roots, there were nine species of medicinal herbs, she pounded them in a mortar and pestle and made a pomace from the herbs that could be pounded and pounded.

Peasant relations are expressed through the ways in which knowledge is passed on, as Mr Lázaro's grandmother did, through family inheritance and also through the way in which they organise space, which, according to Braga, is where the subjects' manifestations take place. According to this author, "Space is seen as man's construction and not as something given; it is his daily life" (BRAGA, 2007, p.69).

He put them in a pot and left it to simmer for about three hours. Then she took the broth off the heat, strained it, added rapadura and returned it to its normal state, but with the herbal tea inside. Every morning, she put him on one side of the door, sat on the other, lit a straw cigarette, gave him a piece of that rapadura to eat and kept watch.

Mr Lazarus was only twelve years old, but he was a child who still didn't understand and, in fact, this event remained engraved in his memory. His mother understood a little too, but he didn't learn from either of them.

That's how it started. He had always grown up with an interest in roots, but it was only after a long time that he went on to work as a root grower and, soon afterwards, as a holistic therapist. He took a few courses in this area, but that's where his desire to work with medicinal herbs came from. His interest really came from his grandmother's help in healing his wound with herbal medicine.

Over the years, in 1998, he started working with the plants, influenced by someone who came from Goiânia-GO to Quirinópolis for the livestock festivities. This increased his desire to work with plants, which continues to this day with his treatments. From 1999 to 2000, he started working for a company that made treatments with natural products imported from the United States, called "multi-level marco". He never left his root clinic, even during periods of success in the business or the difficult times that life throws at him, he says.

Imported products used to enter Brazil through an ordinance system. However, this document expired and they were unable to renew it. As a result, there were shortages of some products and the landmark chains declined. The idea of forming distributors in holistic therapy arose with the aim of increasing sales of the remaining natural products, but it was even worse, because, according to the interviewee:

> Because in these phytotherapy/holistic therapy/naturopathy courses, we also learnt about phytotherapy, which gave emphasis and structure to what we were already doing, which was working with raisins, and it got worse when we gave up.
> Even once the multilevel framework was in place and I firmly took root in the raizada, right, in working with medicinal herbs. (Verbal information)[5]

From then on, Mr Lázaro specialised in medicinal herbs and also in Holistic Therapy, which he learned in the courses he took at[6] . He was a holistic therapist for eleven years, but with the loss of his first wife, with whom he lived for 25 years, he was no longer able to work as a therapist, and continued with his roots business, which he still runs today (2015).

In 2003, there was an opportunity to develop a phytotherapeutic product that could help producers in agriculture. Activities with this product came to a standstill until September 2015, when, according to the interviewee, "God sent a mate from São Paulo who came to reactivate the product again". So there is still hope of working with this product and, if the result is positive, he probably won't be able to work with medicinal herbs any more because of the activities he will have with this new product. But,

[5] Interview conducted in October 2015 by the author at the Quirinópolis Free Market.

[6] In these therapy and naturopathy treatments, Mr Lázaro and his first wife Dona Nelcy (in memoriam) worked together, but they had already decided that they were going to leave this job and just stay with the roots stall, since Dona Nelcy was involved in Mr Lázaro's entire career, from collecting the roots to selling them at the market, and so they would only finish with the patients who had already paid for the therapies and those who had returned. However, a month before the deadline they had set, Mr Lázaro's wife died and he decided to bring the deal forward. So he stopped the therapies once and for all, although he continues with the activities of the root bank.

according to him, until it is ready, he will continue to work with the stall and the medicinal herbs at the Quirinópolis street market.

Because he is very religious, during the interviews he always remembers that God has blessed him and maybe that's because he has always helped many people. He tells them "it's not me who is good, but God".

After converting to the evangelical faith, in all his prayers, before starting or even after work, he puts himself in God's hands and this, according to him, has helped him a lot, a practice he extends to the people who buy his products.

The relationship between Mr Lázaro and his consumers has always been very important, as they believe in the healing power of the plants and also in his know-how, and in fact this relationship often comes from many years of treatments and medicines that they have used, always with positive results.

In the research space used, two people were interviewed who deserve to be highlighted. They were effective, long-term treatments. One of the interviewees became very emotional as he told us about his treatment. Everyone answered the questions with satisfaction, telling of the cure they had obtained and also of the treatment that Mr Lazarus had provided.

Interviewee A, who has been using home remedies for around 40 years, has completed higher education, lives in the city centre and has been a customer of Senhor Lázaro's for 12 years, says that when her daughter was very small, she had many health problems, such as reflux, lung problems and a lot of diarrhoea. She was very weak and Mr Lázaro treated her for two years with roots and bottles. Everything he prescribed was with roots or leaves to make tea, but there were some pills from folk medicine.

Years later, she contracted health problems and was also very weak. She used prescription drugs until she decided to stop taking them and start a treatment with Mr Lazarus, because instead of helping her get better, they were making her worse and worse. So he treated her with clay for thyroid and gastritis problems, many natural medicines with roots, bottles and natural pills. Today, she no longer uses chemical remedies.

The interviewee said that she is also concerned about the extinction of folk medicine, and that for her, these natural remedies are a way of carrying out a treatment and obtaining a cure in a natural way at a very low cost. They were long treatments, and today she is very grateful to Mr Lazarus for having helped her cure her illness.

Interviewee B is a 59-year-old consumer, who is also Catholic, has a university degree and a master's degree, lives in the city centre and has been a customer of Senhor Lázaro for 20 years. He told us that his wife had a health problem with her uterus, as she had a myoma and several cysts on her ovaries. He heard about Mr Lázaro, from whom he bought a bottle for his wife to treat the illness. Soon after the treatment, it was realised that she no longer needed to have her uterus and ovaries removed.

Nowadays, this interviewee uses folk medicine much more, and recommends that people who use this medicine need to have patience and perseverance, because it doesn't harm the body and you have to do the treatments over the long term. This way, the problems will gradually improve. However, he warns that you don't have to forget what the doctor prescribes, because it's an aid to the other medicine and you have to be persistent, because only God can work miracles.

For him, folk medicine tends to become an even more common practice, but his only concern is the devastation of the natural vegetation, because he believes that the cure is in the Cerrado and, according to information he has acquired, out of every ten plants in the Cerrado, eight are medicinal. There is therefore great concern about the extinction of the Cerrado.

He also emphasises the possibility of setting up a shop selling natural products in our town, but for the time being people look for Mr Lázaro at the market, where he has a wide variety of roots and other parts of plants that help with treatments. It's very important, because there are people who arrive weakened, with certain illnesses, and these roots give them renewed hope for a cure.

Finally, this interviewee continues to switch from pharmaceutical remedies to folk medicine. According to him, the older he gets, he will look for remedies that don't harm his body. He says that taking a pathological remedy cures one problem, but others arise. For him, folk medicine is linked to faith, because if you use medicinal herbs but don't believe and have faith that it will work, you won't get the expected cure.

Final considerations

The culture of the raizeiro is, in fact, close to extinction due to the deforestation of the Cerrado. As a result, this know-how with natural remedies could be forgotten as the years go by, in other words, there is a risk of losing the knowledge acquired over generations, inherited from ancestors who left great legacies for the organisation of society and space.

In this way, the Cerrado is important to the population, who depend on it for their socio-cultural relations, especially when it comes to gathering food and medicines. In this sense, many raizeiros, like Mr Lázaro, live in urban areas but still depend on the countryside to carry out their activities, a legacy of peasant traditions for making home-made medicines.

Because of this, Mr Lázaro is very concerned about the continuity of this culture of folk medicine in the near future, because with the practice of deforestation, there is a clear threat of the extinction of this popular wisdom.

He may not realise the loss of this knowledge, but it is certain that these changes in ecosystems are leading to the disappearance of medicinal herbs.

The Feira Livre in Quirinópolis has been a fundamental space for the preservation of popular memory and is of great importance to market traders as a support for peasant family farmers, as they not only sell their produce, but also buy goods from other producers that they don't produce themselves.

This research, based on studies carried out at the Quirinópolis Free Trade Fair, indicated that both Mr Lázaro and his consumers feel threatened by deforestation and that, for them, the know-how of the raizeiro is very important to society, which emphasises an immense desire for this profession to be valued. According to them, a specific space for this practice of natural products would be essential.

It should be emphasised that, despite all the setbacks, home remedies, produced naturally and without preservatives, are still present in our local and regional society. According to Mr Lázaro, if there were incentives to build medicinal plant gardens, this profession would be better regarded by society and the population would have access to cultivation as well as knowledge and know-how.

Finally, the specificities and singularities of the raizeiro and their work with folk medicine are highlighted, testifying to the fact that the use of home remedies is still present in our society. In order to maintain the practices of the people of the Cerrado, and especially the raizeiros, it is necessary to implement public policies to encourage and protect not only the know-how of traditional populations, but also measures to intensify nature conservation, to guarantee areas for the collection of medicinal plants, which are so important to society, as has been demonstrated throughout history.

References

BORGES, Viviane Custódia. Pequi, Jatobá, Algodãozinho... The biodiversity of the Cerrado in popular medicine. Goiânia, 2011. Available at: <projetos.extras.ufg.br/posgeo/wp-content/uploads/.../Viviane-Tese-Final.pdf>. Accessed on: 16 April 2015.

. The "Cerrado de Pé": the potential of medicinal plants. Goiânia, 2013. Available at: <http://www.revistas.ufg.br/index.php/atelie/ article /view/9224/13923>. Accessed on: 16 Apr. 2015.

DOLLFUS, Olivier. Geographical space. 4. ed. São Paulo: Difel, 1982.

PEREIRA, Antônio Mozair; GARCIA, Edilene Maria; OLIVEIRA, Marcos Gondim de. Socio-economic profile of market traders in Quirinópolis (2009). 67f Monograph (Graduation) - State University of Goiás, Quirinópolis, UEG, 2009.

HORTA, Maria de Lourdes Parreiras; GRUNBERG, Evelina; MONTEIRO, Adriane Queiroz. Basic guide to heritage education. Brasília: National Historical and Artistic Heritage Institute, Imperial Museum, 1999.

MORAES, Antônio Carlos Robert; COSTA, Wandeley Messias da. Geografia Crítica: a valorização do espaço. 2. ed. São Paulo: Hucitec, 1984.

PEREIRA, Antônio Mozair; GARCIA, Edilene Maria; OLIVEIRA, Marcos Gondim de. Socio-economic profile of market traders in Quirinópolis. 2009. 67 f. Course Conclusion Work (Graduation) - State University of Goiás, Quirinópolis, 2009.

SANTOS, Milton. Metamorphoses of inhabited space. São Paulo: Hucitec, 1988.

SCARABELOT, Maristela; SCHNEIDER, Sérgio. Short agri-food chains and local development - a case study in the municipality of Nova Veneza/SC. s.l: s.e., 2012.

CHAPTER 2

CRAFTS AND ARTISANS IN THE URBAN AREA OF QUIRINÓPOLIS: a way of working with meaning
work with meanings that seek means to achieve a certain end

SILVA, Marta Corrêa da[7]
SANTOS, Jean Carlos Vieira[8]

Introduction

The cultural identity of the population of Quirinópolis is associated with the agricultural and livestock activities developed in the Cerrado biome and these rural segments have contributed to the emergence of other activities that are less economically representative than farming, but which have been fundamental to maintaining the way of life in rural and urban homes in the interior of Goiás. One of these activities is handicrafts, also understood here as folk art, know-how and manual labour, since no work developed by the hands of local people is the same as in industrial processes.

It is therefore possible to state that the aim of this work is to diagnose, get to know and present the handicrafts/manual labour of the city of Quirinópolis, based on pieces that were initially made for the domestic use of residents and which have now taken over some of the city's commercial premises. The main handicrafts found in this urban area are crochet, embroidery, rugs made from cord and patchwork and gourds with paintings and drawings.

With this in mind, we set out to analyse and find answers to the following questions: What handicrafts are produced? What are the places where this know-how is sold? Who are the people who, with their creativity, reinvent an art with particular and popular traits?

The subject of this article is the urban municipality of Quirinópolis, located in the microregion of the same name in the state of Goiás. In this respect, this work is justified by the lack of geographical literature and literature from related areas, such as tourism and economics, addressing the themes of manual labour, handicraft know-how and artisans in Quirinópolis. In order to achieve the proposed objectives, the research starts with a discussion of the geographical category of **place**, arriving at the commercial and

[7] Graduated in Geography at the State University of Goiás, Quirinópolis Campus. .martacs.10@hotmail.com
[8] Professor of Geography at the State University of Goiás, Quirinópolis Campus. .svcjean@yahoo.com.br

creative territories of the subjects involved in handicraft know-how.

Theoretical references (books, newspapers, the internet and magazines), fieldwork, photographic surveys, informal interviews, questionnaires and office work were fundamental to the construction of this work. Fieldwork is not simply a time to take photographs, conduct interviews or apply questionnaires. According to Moura and Silva (2009, p.9-10), it must "[...] guarantee interpretative approaches to the reality visualised".

Santos (2010, p.31) helps us to understand the relevance of this practice, because it is in the field that the "[...] gaze goes beyond appearances and seeks to make history with intensity, through detailed and recognisably unique documentation". This necessary journey requires:

> [...] a disciplined posture and an effort of memory, as well as intense vigilance so as not to treat the subjects as exotic. When making notes in the field notebook, it is essential to take into account the emotions of the subjects, as well as the emotions of the person making the notes, because they allow openings, including subjective ones, to understanding reality (SANTOS, 2010, p.33).

Through office activities, the data collected was correlated, interpreted and analysed. It can be said that Quirinópolis, like the whole of the interior of Goiás, has a handicraft with its own particularities and art, very characteristic of its people, its traditionalism, its way of being and thinking and, above all, its uses and customs with traces of the Cerrado.

These, in turn, are truly combined with the various original products from the same land, sold in the state's tourist towns, such as Caldas Novas, Rio Quente, Trindade, Pirenópolis, Lagoa Santa, Goiás (Old Capital) and the capital, Goiânia.

The art of embroidery, creating clay pieces in the old potteries and reinventing in rural and urban kitchens in this land is ancient. Don't be surprised if we tell you that this same antiquity has accompanied it since the arrival of the first inhabitants, which leads us, in this article, to consider, with a certain degree of certainty, that craftwork was the oldest (if not the first) creative form to flourish in this Cerrado. It's worth emphasising that there is a natural tendency in the region to build different pieces using creative knowledge and hands.

The place of crafts and the craftsmen of the place

Geographical science is an area of knowledge that studies places, territories and artisanal landscapes. It is in this segment of scientific research that it becomes important to discuss the geographical category of place, interconnecting social, economic, cultural and tourist elements, as well as weaving contextualised discussions on place that also refer to the art of hands that create and shape.

The handicrafts that exist in places like the municipality of Mateiros, in the Jalapão (TO), where golden grass is used, have become one of the destination's main tourist attractions and the community's main promoter. In this way, it is presented to tourists and visitors as a means of social integration, cultural inquisition and a source of economic resources for the development and growth of the local economy.

It can be said that the handicrafts developed in the different parts of the Cerrado biome and other Brazilian biomes are a branch of the economy that can showcase the cultural manifestations and traditions of each people, enabling the development of locally-based tourism and, at the same time, generating income for the local artisans. According to Marques (2011, p.50), in these places, a very particular form of "everyday practices" is developed, bringing creative actions and artisanal traits. The author states:

> [...] that these places are, first and foremost, a space. But it differs from it in that it is an environment marked by relationships of identity and belonging. An individual's place is unique, particular and subjective. From this perspective, the constitution of place takes time and depends on experience (MARQUES, 2011, p.70).

In fact, place is "[...] where people's sensitisation and sociability occur, through everyday practices, diverse social networks and ways of life that are key components in regional planning [...]" (SANTOS, 2010, p.244). Silva (2010, p.161) explains that place is the "[...] social, historical and cultural movement that affirms the society/nature relationship in the place. Subjectivity, identity and a sense of belonging are the categories that indicate a place". In this context, it's possible to say that the place of handicrafts is also the artisan's, and can become an attraction for outsiders.

It is from this perspective that Santos (1997) refers to place, saying that it is dialectically superimposed on the axis of succession, which transmits the external times of the higher scales, and the axis of internal times, seen as the axis of coexistence, where everything merges, definitively linking the notions and realities of space and time. For the author, place expresses objective relations in articulation with subjective relations, vertical relations resulting from hegemonic power, imbricated with horizontal relations of coexistence and resistance.

If the place expresses relationships, when imagining the manual labour that leads to the production of handmade or manual pieces, it can be said that the homes or territories where this production takes place can be characterised as a place of sociability which, in the words of Silva (1991, p.31), is "[...] a relationship of simple or complex causality, such as reciprocity and reference". For the author:

> These mediations relate man and place, manifesting themselves as physical complexes of work and energy exchange, in other words, as the set of social and natural relations produced by work, work which is the transformation of reality based on determination, and which implies the passage of energy from one receptor to another by means of a conductor (SILVA, 1991, p.22).

The place of manual or artisanal labour can also be a social territory perceived and understood as a whole and which has social life as its reference point - this whole is seen as an analytical multiplicity of social exchange actions in a heterogeneous reality. For Silva (1991, p.134), the role of social geography is fundamental in this discussion, which aims to "[...] study the social place and its population as a manifestation of social space formation".

The relationships in the social place, seen in terms of the lives of individuals who develop their know-how, can be portrayed as a socio-spatial practice of communities with strong rural and traditional traits. This can be seen in the way people appropriate the materials that the local base can offer, organised by socio-cultural traditions or by the need to produce their own objects. It should be emphasised that these spaces are territories where human life takes place, determining it according to their habits and customs.

According to Santos and Alves (2005), it is in the place that the mosaic of cultures flourishes, demonstrating the diversity and contradiction of traditions, customs, ways of living and doing things. The manifestations of local culture can be seen in the natural and cultural landscapes, in local traditions, customs and habits inherited from socio-cultural formation, a reality found in the interior of Goiás, in the municipality of Quirinópolis and in the Cerrado region. In fact, this landscape can make it possible to get in touch with, recover, appreciate and maintain the values of the place and the region.

In general, the different crafts and handicrafts carried out in the municipality of Quirinópolis take place in places that can be defined as family production units. They are not well known or even studied with a view to assessing their tourism and cultural potential, as well as the possible impacts that could result from public and private initiatives aimed at expanding the municipality's tourism offer and generating income for family craftspeople.

When discussing place and the production of everyday life, Damiani (1999) points out that it is necessary to incorporate into space the critique of everyday life, which emphasises the reproduction of social relations. Everyday life becomes a level:

> [...] of analysing the real, which is important when social reproduction entirely affects the reproduction of life. [...] the ways of life, customs and symbolisms of

> a people cannot be completely erased, no matter how rigorous the reproductive processes involving companies and political strategies. Everyday life is also created by the mode of production, it is not alien or marginal to it, although it is unequal to other sectors, moments and situations of this form of production. Everyday life, as a simulacrum of the full life thus modelled, allows the mode of production to function. [...] everyday life, in relation to the economic and political, expands the universe of analysis to include so many other relationships between individuals and groups, including particular, local ones. It includes what is lived, subjectivity, emotions, habits and behaviour (DAMIANI, 1999, p.162-163).

For this author, the place defined by everyday life is the whole of society, not just its economic and political aspects. Therefore, the places of manual/artisanal labour in Quirinópolis are part of the daily lives of people with their artistic skills and sensibilities. In this vein, Carlos (1999, p.182) points out that living in a place reveals itself as the constitution of a multiplicity of social relationships, with a spatial practice "[...] that is at the basis of the process of constituting identity with the place and with the other and that escapes the hegemonic homogenising rationality (imposed by the consumer society), which defines a time and a mode of use".

In this case, it can be said that in craft production sites there is a sociability marked by activities developed in a small-scale production logic, taking place in different neighbourhoods of the city and often in neighbourhood relations. This production is commercialised in the homes themselves or in public spaces:

> [...] a multiplicity of meanings for society according to culture, habits and customs. [...] the place of communication, dialogue, the extension of living, where children can play and everyone can expose themselves, talk and make claims. Here, visibility is produced that creates identities - the identity that humanises relationships through the bonds of coexistence and the feeling of belonging to the group and the place (CARLOS, 1999, p.184).

The use of public places by the craft trade as a form of appropriation is realised as an expression of a traditional mode of production, in other words, the space is experienced by the artisan using their senses, sensibilities and creativity, which form the practical basis of their perception of the outside world. In this vein, in the study proposed here, it is important to think about craft production sites, whether tourist sites or not, from the point of view of their meanings and what they represent for societies living on the margins of industrial and capitalist processes.

Handicrafts: a theoretical reflection

When it comes to handicraft work, especially the know-how or manual labour born in small towns or rural areas, it can be said that this labour is part of a spontaneous popular culture, as it is learned in family life. In this context, we can think that this knowledge is the result of a tradition in the cities of the interior of Goiás, being the behaviour of individuals who bring with them skills and creativity based on nature and the existing way of life.

In this way, we can think of handicrafts as a form of work with meanings that seek means to achieve a certain end, giving shape and meaning to a certain popular culture that is both tourist and non-touristy. In this context, Claval (2006) argues that man apprehends the world through his senses: he observes shapes, hears noises and smells what surrounds him.

The movements of your body are a direct experience of space. When you eat or drink, taste reveals other properties of the world around you. It can therefore be said, based on Claval (2006), that manual or artisanal labour is an indication that subjects receive from their senses. In this case, it can be said that

> [...] the information that makes up cultures passes endlessly from individual to individual. It passes from one generation to the next, so that society remains even if its old people disappear and are replaced by young people. Information circulates between neighbours, between friends, between work or business partners. In the course of these exchanges, everyone receives know-how, knowledge and discovers attitudes and beliefs that were foreign to them; a more or less large part is retained and internalised. The content of the messages exchanged cannot generally be understood outside the context in which the partners meet (CLAVAL, 2006, p.95).

Based on Santos (2011), handicrafts can be understood as the technique and work of the artisan, an art of making objects, while the artisan is the craftsman or subject who makes the handicraft. The craftsman is the most finished product of what the human imagination can do in the field of manual labour: everything produced by man in this field has art; even what he himself, in his aesthetic and over-refined demands, wants to understand that it is not.

Since time immemorial, human beings have been spiritually fulfilled by making objects for personal use and utensils that their daily lives have demanded and perfected. So:

> Man, by nature and analysed in general terms, is a craftsman, an artist, given his ability to reconcile and adapt his thoughts to the correct and harmonious movements of his hands. He possesses imagination, taste and manual skill (SANTOS, 2011, p.471).

Handicrafts are the purest manifestation of popular art, carried out by skilful hands, almost always unknown, "[...] who operate in accordance with the taste and intuition of the born artist, whose learning and maturity they pick up in the school of life, where the student, as a general rule and as a paradox, is the teacher of himself" (SANTOS,

2011, p.471). This way of characterising handicraft work is no different from the realities encountered during the fieldwork in the city of Quirinópolis.

In these terms, the preservation of artisanal work or manual labour in places promotes "[...] the maintenance of the community's living history" (CORIOLANO et al., 2009, p.153) and can contribute to the development of locally-based tourism, becoming an income-generating art or an income supplement for families or groups organised around this activity.

According to Bezerra (2013, p.118), handicrafts are "[...] a symbol of a community's social and cultural representation". Therefore, in order to highlight these social representations, it is necessary to:

> [...] to feel and let oneself be invaded by the feeling of discovery. It is a singular imaginary universe, a reflection of a unique social scenario, culturally constructed over many decades as a representation of a people, of a different but original society (BEZERRA, 2013, p.125).

It can be said that researching the handicrafts or simply the manual labour of Quirinópolis is to bring a sense of discovery that represents a people whose traditions are based on religion, agriculture and livestock farming. It is possible to refer to this manual labour in the daily lives of people in their homes or small businesses as a cultural practice or a habit to pass the time, in other words, handicrafts or manual labour are part of the culture of Quirinópolis. It wouldn't be daring to say, in this article, that it's a heritage of the place:

> [...] culture is made up of the mediating set of instruments (utensils and other materials) and the framework of norms that govern the various social groups (ideas, crafts, beliefs and customs), all of which come from the signs created by man. When considering culture, whether simple or complex, we will always be in the presence of a vast material and sign apparatus (ZUIN; ZUIN, 2009, p.42).

Together with handicrafts, different cultures are born that help people pass on historically constructed knowledge. This knowledge seeks to preserve the particular history of a family, a generation, a town, a farm, a region and a country. It can be said that craft practices and experiences developed in different contexts and locations across the nation:

> [...] have brought the answers sought and demonstrated surprising results. The first of these can be seen as a new vision of Brazilian cultural heritage in its diversity of manifestations, tangible and intangible, consecrated and non-consecrated, as a primary source of knowledge and learning, to be used and explored in the education of children and adults, as part of the curricula and subjects of the formal education system, or as an instrument of individual and collective motivation, for the practice of citizenship, the recovery of the self-esteem of cultural groups, and the establishment of an enriching dialogue between generations (HORTA, 1999, p.5).

By appropriating the discussion presented, it can be said that craftwork is a cultural phenomenon (and asset) in places throughout the country, not just in the cities of inland Goiás. It can be transformed into understanding, internalisation and appreciation, in other words, into a method of heritage education as a continuous process of experimentation and discovery.

Its richness and potential can only be assessed and evaluated by those who experience it in its various fields of heritage, education and community action. Craftsmanship is a craft:

> [...] cultural heritage as a primary source of knowledge and individual and collective enrichment. Critical knowledge and conscious appropriation by communities of their heritage are indispensable factors in the process of sustainable preservation of these assets, as well as in strengthening feelings of identity and citizenship (HORTA, 1999, p.6).

All the actions through which people express their specific handicrafts constitute and form part of their culture which, over time, acquires different forms and expressions, often becoming elements of their own.

development of tourist sites. For Horta (1999, p.7), this culture is an eminently dynamic process, "[...] passed down from generation to generation, learnt from ancestors and created and recreated in the daily life of the present, in the solution of small and large problems faced by each society or individual".

Also according to Horta (1999), it is important to recognise that all peoples produce culture and that each has a different way of expressing and accepting it. There are other forms of cultural expression that make up the living heritage of Brazilian society: handicrafts, ways of fishing, hunting, planting, cultivating and harvesting; using plants for food and medicine; building houses; cooking, dances and music; ways of dressing and speaking, religious and popular rituals and festivals, social and family relationships - all reveal the multiple aspects that the living and present culture of a community can take on.

Getting to know the craftswomen of Quirinópolis

Before presenting the social and economic profile of the craftswomen of Quirinópolis, it can be said that public and private spaces, such as homes, squares, streets, leisure areas, pavements, public and private buildings in the city, have been the places where the crafts produced in the urban environment have been commercialised. All the handicraft work carried out in these places follows a certain ritual, ranging from individual concentration and group conversations to the final product in embroidery, carpets and paintings.

Craft work is "[...] knowledge - imaginary/real - sometimes transmuted into the irrational in practice, by the new generation in fact, what exists in this craft is the "[...] sensitivity to understand life and the universe through the imagination that comes out of the heart and takes on a real form in everyday practice" (BEZERRA, 2013, p.129-130).

It can be seen that craft know-how, not only in the municipality of Quirinópolis, but also in other regions of Goiás and Brazil, presents the "[...] integration of people with each other and with the place, creating a very strong identity" (CORIOLANO et *al.*, 2009, p.45). In this context, the craft places in the urban space of Quirinópolis will be analysed in this work in a comprehensive way, making a diagnosis of the reality found, because, as Santos (1997) states, such a place constitutes the dimension of existence that manifests itself from a daily life shared between the most diverse people who are the basis of the common life that is sometimes provided by craft work, becoming part of the culture of the place.

In this case, Claval (1999, p.11) reports that culture "[...] is a common field for the human sciences as a whole"; therefore, the geographer's gaze becomes relevant to understanding Quirinopolino's handicrafts, without dissociating them from their subjects. Rosendahl and Corrêa (2003, p.9) also point out that "[...] cultural geography is a significant subfield of geography, which spread from Europe and is already a century old. However, cultural geography is not yet as important in Brazil as it is in the USA and Europe".

According to Rosendahl and Corrêa (2003), Brazil's cultural heterogeneity, the result of long, complex and spatially differentiated processes involving society and nature, makes it an excellent field for cultural geography studies, but this research focussed mainly on traditional societies, with little reference to urban-industrial societies .

In Quirinópolis, artisans sell their art at events held in the reception area of the city's theatre, a public space "[...] considered one of the municipality's main architectural works, built at the end of the 1980s" (SANTOS, 2010, p.202). Trade also takes place in the homes of the artisans themselves, and this reality culturally integrates the population of Quirinopolina with the know-how of these anonymous people, incorporating elements of the place into this context, such as trade with neighbours, which expresses relationships of proximity, ways of life and values that shape families in the interior of Goiás.

Handicrafts were once sold during the Chica Doida gastronomic festivals[9] , but this was cancelled due to local political differences. Logically, the biggest loser was the local population. Santos (2010) highlights the presence of handicrafts at the local gastronomic festival, showing them to be an attraction for those who visit the event:

> Various regional products are sold during the Chica Doida gastronomic festival, including green corn flour, which has been produced by hand for over 30 years by a family at Fazenda Fortaleza do Castelo [...]. Another attraction of the event was the **"Festival Warehouse", where it was** possible to buy manioc flour, rapaduras, sweets, cheeses, curd cheese and other rural delicacies, as well as local handicrafts (SANTOS, 2010, p.217-218).

The experiences and examples presented show the structural transformations of spaces in a small town into places to sell handicrafts. Another place of sale and exhibition is a shop on the corner of the town's main road (Avenida Brasil) and the Church of Nossa Senhora da Abadia (Matrix and Patron Saint), located in the centre of the urban core. The shop functions as a bar and restaurant at night and during the day the place is used to sell local handicrafts.

At events such as the patron saint's feast (Nossa Senhora da Abadia), the Livestock Festival and the open-air fairs at Praça do Circo (Wednesday) and Feira Coberta (Sunday), you can also find the handiwork of local residents. According to Martins (2015, p.93), "[...] street markets are present in modern times and perhaps, depending on the region to be analysed, become one of the most attractive activities for consumption and leisure". In towns in the interior of Goiás, whether they are tourist towns or not, the use of the space by their artisans can be seen in these open-air markets, selling their produce to locals or tourists.

Another relevant point in this approach is the Youth of the Future Social Complex, a public project to support manual and artisanal work in the city, whose headquarters are located on Avenida Dom Pedro I, in the city centre, and which receives support from the Municipal Department of Promotion and Social Assistance. The Young People of the Future project concentrates its activities in four centres: pedagogical, sporting, cultural and professional qualification. The cultural centre stimulates cognitive, bodily and artistic development, awakening creativity and cultural production (QUIRINÓPOLIS, 2014).

---

[9] Mrs Petronilha Ferreira Cabral and her husband João Batista da Rocha were responsible for creating the Goiás dish (around 1958), as they liked to gather family and friends for pamonhadas. The story goes that, at one of these pamonhadas, they ran out of corn husks and, as there was still a lot of dough, they decided to invent a dish. In the context of regional literature, **"[...] they created the recipe,** adding cheese, sausage, jiló, onion and other condiments to the already seasoned dough and **baked it in the oven" (SANTOS, 2010, p.216).**

Among the courses on offer at this centre are: hand embroidery, machine embroidery, **biscuit**, crochet, stonework, the production of handmade vases, handmade floor coverings, among others, all of which contribute to the maintenance of locally-based manual and handmade work. Another noteworthy social project in the city is the Association of Mocinhas and Mocinhos de Ontem (AMMO), which provides cultural support to people in the city, focusing on the importance of preserving the past, especially the practice of handicraft work (QUIRINÓPOLIS, 2014).

In order to get to know the target population of this study and provide information about the social group we defined as craftswomen in Quirinópolis, questionnaires were administered to 40 people. Of this group, 15 are from the Young People of the Future project, five from AMMO and 20 work informally from home. This social class can be defined as "[...] popular artisans, whose work has a completely personal style, thus showing their own vision of the world" (WEISS, 1981, p.40).

It is important to emphasise that 100% of the group interviewed is female, as there were no male artisans. Figure 1 shows that 25% of the people working with handicrafts in Quirinópolis are between 41 and 50 years old and 32% are between 51 and 60 years old.

Figura 1 - Age of Quirinópolis' craftswomen

Source: Prepared by the authors.

Another important piece of information gathered during the field research concerns the level of education of the people who work with handicrafts. The results show that 28 per cent of respondents have primary school education, 42 per cent have secondary school education and 27 per cent have higher education, with only 3 per cent not answering this question. In an informal statement, the group with higher education emphasised that the course was made possible mainly by the existence of the State University of Goiás (UEG) in the city, as it is public and free. Thus, the result shows a good level of education among the respondents, which could lead to better organisation for these individuals in the future.

In order to reinforce the craftswomen's professional training, we tried to find out how they got into craft work. The results show that the majority (44 per cent) started working as a distraction or to pass the time; 16 per cent said that they really enjoy what they do and are in awe when they see the finished craft; 37 per cent said that the main reason was to increase their family income and 3 per cent like to give gifts to people and are not looking to make a profit - for them, it's just an art.

Figure 2 shows which religious group each respondent says they belong to. The majority, 67%, say they belong to the Catholic religion, a predominant religious group in the cities of the interior of Goiás and among the residents of Quirinópolis, who have Our Lady of Abadia as their patron saint. Catholic churches are common in the city's various neighbourhoods.

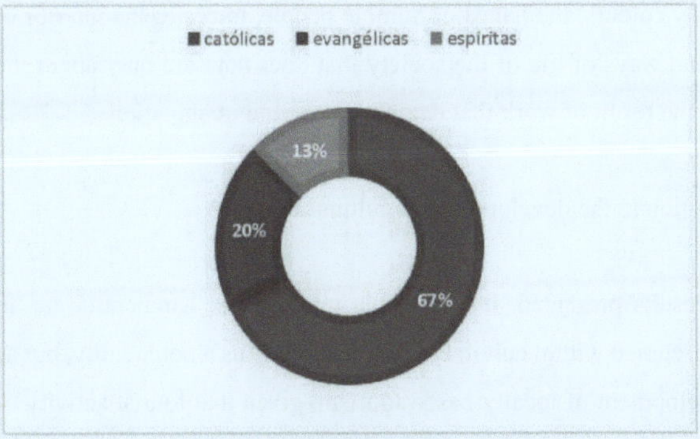

Figure 2 - Religion of the Quirinópolis craftswomen.

Source: Prepared by the authors.

According to the figures collected in the fieldwork, it is possible to highlight that 89% of the respondents do not advertise their handicrafts in any way, 7% do "word of mouth" and only 4% use the internet to publicise their work, especially on their Facebook page and by email. Despite the high number of artisans with a good level of schooling, there is no organised process for marketing their handicrafts.

With regard to selling handicrafts to residents of other towns, 82% said they don't sell what they produce, 3% didn't answer; however, 15% of those surveyed sell to other towns, mainly to people they know (relatives and friends) who live in urban areas in the region, in Goiás and the neighbouring state of Minas Gerais. For this place, the craft activity represents a socio-economic alternative, in terms of citizenship, a source of income and cultural and tourist promotion.

For 94% of those surveyed, manual/artisanal work receives support from the local government (mayor), mainly through social projects and the organisation of events in the city, but there is no project to fund art and manual work in the municipality. Still on the subject of the support they receive, only 32 per cent feel that the town council is present. For 55% of the artisans, handicraft work is a potential tourist attraction because, according to them, people admire the work they do.

Among the main handicraft products made by Quirinópolis artisans are: crochet, embroidery, cross stitch, sianinha, stonework, work with recyclable products, painting, macramé, string and patchwork rugs, knitting, patchwork, free stitch, sewing, painting on gourds native to the Cerrado, biscuit, soaps, vagonite and machine embroidery .

In view of the above, it is important to highlight that the place of handicrafts in Quirinópolis is defined by colours, the hands of humble people, traces of the interior of the Cerrado, daily life and ways of life of the society that does not care only about the profit margin, but about the result of work that demands time and imagination.

Handicrafts as a contribution to the development of cultural tourism

In view of the results presented, it is possible to state that handicrafts are an activity that should be discussed within cultural tourism, not only as a potentiality, but as a possibility for the development of locally-based tourism, given that tourist activity or simply "[...] tourism is always a cultural act, of whatever kind [...]" (PIRES, 2001, p.69).

Historically, tourism has been responsible for spreading various arts, know-how and crafts throughout the Brazilian countryside. It is therefore important to emphasise that culture has always been one of the most important factors:

> [...] the development of tourism is becoming increasingly pronounced: a large proportion of trips are made to destinations that have remarkable cultural factors, such as historical sites, monuments, archaeological centres, pilgrimage centres, ethnic concentrations and many others. Some of these factors can be created artificially anywhere (museums, for example), but others are deeply linked to the places where they developed (historical sites or archaeological centres, for example) and others are linked to the way of life of each people (traditions) (CUNHA, 2001, p.122).

In this way, culture and cultural goods exert a strong attraction and can, by themselves, give rise to the creation of tourist destinations. According to Bezerra (2013), with the development of tourism, inland communities feel valued and, when handicrafts win the favour of visitors, the possibility arises of generating employment and income for families. Thus, craft work becomes "[...] responsible for the strong socio-economic impact [...]. The opportunity to obtain a stable source of income and improve living conditions" (BEZERRA, 2013, p.128) sharpens the wisdom of the locals who make a living from handicrafts.

The art and cultural manifestations of people in small inland towns, like Quirinópolis, can benefit from any municipal tourism programme aimed at incorporating local cultural heritage. In order to take these potentialities into account, even if they are residual, it is essential to carry out public projects that can locate, map and analyse the most potential tourist sites for craft production in the municipality. This will give the local government the opportunity to indicate measures for the development and incorporation of the municipality's artisan communities into tourism, in which this cultural heritage is valued, preventing it from disappearing over time.

In this reflection on handicrafts or manual labour as a tourism product, it's important to think about the places of handicraft work as a tourism offer, because, according to Melazo and Santos (2004, p.186), we must understand that:

> [...] in order to remain a tourist attraction, it is not enough for a place to possess only natural and cultural resources. Tourism needs to work together with other fields of knowledge. The compatibility of the natural and cultural environments is often overlooked in tourism planning.

Handicrafts as a tourist product, if well planned, can become a cultural exchange between locals and visitors. According to Coriolano (1998, p.114), "[...] the organisation of life generating social practices gives each place its own distinctive arrangement. It's as if the place had its own unmistakable physiognomy".

According to the aforementioned author, in tourism, the specific identities of each place are the main local values. Those places that know how to value their memory and culture will enrich tourism; hence the need to preserve knowledge and skills such as the handicrafts found in towns in the interior of Brazil.

In this case, it is understood that the handicrafts of Quirinópolis are a potential tourist attraction for the municipality, which could promote sustainable development in urban family production units, generating income and labour, by considering know-how as a possibility for developing the quality of life and improving the lives of local populations. The idea is to show craft work as a means for local development, as this would increase the quality of life of family producers and even reduce problems related to unemployment. From this analysis, it can be said that the:

> [...] cultural tourism in its broadest sense gives a new role to environmental and cultural resources in the localisation of tourism on a national and international scale [...]. Some monuments, historical-artistic sites and cities have been visited more than certain coastal resorts (VERA et al., 1997, p.69).

Handicrafts are a cultural tourism product resulting from popular ideas, habits and beliefs, produced through the production of material artefacts and without the use of machinery, including the landscape and the local environment as a source of inspiration. This product is socially defined by the communities responsible for its existence, expressing the lives of social groups articulated in their territories, as well as their ideas and values.

Final considerations

It is hoped that this research has succeeded in presenting the handicrafts of Quirinópolis as a possibility for the study of cultural geography and the geography of tourism, which is now another branch of geographical work. It encompasses analyses of everyday objects, the representation of nature in art, studies of the meaning of landscapes and the social construction of identities based on places where craft production takes place. Our research focus does not include a theoretical investigation to differentiate concepts of craft work, manual labour and popular know-how, but rather the demonstration of a local culture, with its daily life and social customs approached from a series of theoretical perspectives.

In view of this, it should be noted that the place of handicrafts in Quirinópolis is defined by the daily life and way of life of an urban and rural society, in which the economic and political aspects are the least relevant. This everyday reality is in place and constitutes a particularity of the municipality, made up of facts and situations that keep life going. In this way, the history of artisanal know-how begins in the locality, involving close, ordinary and unique relationships. The most intimate everyday life, at the same time, situates its place in society, in the daily life of the place, signifying a unique specific area, identified as such by the manual labour of crochet, embroidery and the art of cooking.

These places are not seen as "data" or "tourist potential", since the municipal priority is to concentrate its actions on the development of agribusiness, and not on cultural tourism - it's a mental reality impregnated in the logic of local government.

Crafts are not valued if we analyse the concrete actions of the local market, and it is possible to say that, in this place, this activity is considered generic, as it is not seen as important by the private and public logics. In fact, handicrafts cannot be understood merely through the observation and will of the artisan; therefore, other sectors need to be aware of this legacy of the past and present.

References

BEZERRA, N. A. P. Representations of the environment in the imagination of Capim Dourado artisans in the municipality of Mateiros - TO. In: CORCINIO JÚNIOR, G. F.; SILVA, V. C. P. da. Nature and imaginary representations. Curitiba: Appris, 2013.

CARLOS, A. F. A. The consumption of space. In: CARLOS, A. F. A. (Org.). Novos caminhos da Geografia. São Paulo: Contexto, 1999.

CLAVAL, P. Cultural geography. Trad. Luiz Fugazzola Pimenta and Margareth de Castro Afeche Pimenta. Florianópolis: Editora da UFSC, 1999.

. Approaches to cultural geography. In: CASTRO, Iná Elias de; GOMES, Paulo Oesar da Costa; CORRÊA, Roberto Lobato (Orgs.). **Geographical explorations:** journeys at the end of the century. Rio de Janeiro: Bertrand Brasil, 2006.

CORIOLANO, L. N. M. T. Do local ao global: o turismo litorâneo cearense. Campinas: Papirus, 1998.

A ARAÚJO, A. M. M.; VASCONCELOS, F. P.; ALMEIDA. Local productive arrangements for community tourism: actors and changing scenarios. Fortaleza: EDUECE, 2009.

CUNHA, Licínio. Introduction to tourism. Lisbon: Editorial Verbo, 2001.

DAMIANI, A. L. Place and the production of everyday life. In: CARLOS, A. F. A. (Org.). Novos caminhos da Geografia. São Paulo: Contexto, 1999.

HORTA, Maria de Lourdes Parreiras. Basic guide to heritage education. Brasília: IPHAN; Imperial Museum, 1999.

MARQUES, Luana M. A festa em nós: fluxos, coexistências e fé em Santos Reis no Distrito de Martinésia - Uberlândia (MG). 2011. Dissertation (Master's in Geography) - Federal University of Uberlândia, Uberlândia, 2011.

MARTINS, Pollyany Pereira. The socio-spatial dynamics of Aragarças, Goiás: everyday life in the construction and structuring of urban space. 2015. Dissertation (Master's in Geography) - Federal University of Goiás, Jataí, 2015.

MELAZO G.; SANTOS, R. J. Event tourism and the cultural tourism potential of Uberlândia (MG), the largest urban centre in the Araguari River Basin. In: LIMA, S. do C.; SANTOS, R. J. Gestão ambiental da Bacia do Rio Araguari. Uberlândia: UFU; Brasília: CNPQ, 2004.

MOURA, Poliana S.; SILVA, Mirella L. Fieldwork in the tourist landscapes of Destination Canastra - Minas Gerais. In: SANTOS, Jean, C. V. (Org.). Landscapes and tourist destinations in geographical research. Uberlândia: Composer, 2009.

PIRES, M. J. Lazer e turismo cultural. Barueri: Manole, 2001.

QUIRINÓPOLIS. Casa Cor magazine dedicates two pages to the Jovem do Futuro social project. 2014. Available at: <http://www.quirinopolis.go.gov.br/estrutura/galeria/janela.php?id=1145>. Accessed on: 15 August 2014.

ROSENDAHL, Zeny; CORRÊA, Roberto Lobato. Introduction to cultural geography. Rio de Janeiro: Bertrand Brasil, 2003.

SANTOS, Jean Carlos Vieira. Regionalisation policies and the creation of tourist destinations between Lago de São Simão and Lagoa Santa in Baixo Paranaíba Goiano. 2010. 367 f. Thesis (PhD in Geography) - Institute of Geography, Federal University of Uberlândia, Uberlândia, 2010.

SANTOS, M. A Natureza do espaço: técnica e tempo, razão e emoção. 2. ed. São Paulo: Hucitec, 1997.

SANTOS, R. Monteiro. History of the Municipality of Lagoa. Lisbon: Colibri, 2011.

SANTOS, R. J.; ALVES, K. B. Register of the cultural and built heritage of the areas directly affected, surrounding and influenced by the Capim Branco I and II Hydroelectric Power Plants. Uberlândia: Composer, 2005.

SILVA, Armando Corrêa da. Geography and social place. São Paulo: Contexto, 1991.

SILVA, Lorranne Gomes da. Construction of place: trajectories of the Avá-Canoeiro in the Cerrado of northern Goiás. In: PELÁ, M.; CASTILHO, D. Cerrados: perspectivas e olhares. Goiânia: Vieira, 2010.

VERA, J. F.; PALOMEQUE, F. L.; MARCHENA, M. J.; ANFON, S. Análisis territorial del turismo. Barcelona: Ariel, 1997.

WEISS, Loise. Playful activities with junk. São Paulo: Scipione, 1981.

ZUIN, Poliana Bruno; ZUIN, Fernando Soares. Tradition and food. Aparecida: Ideias & Letras, 2009.

CHAPTER 3

PRACTICE, SPATIALISATION AND PROFILE OF SKATEBOARDERS IN THE CITY OF CITY OF QUIRINÓPOLIS/GO IN 2014 AND 2015[10]

DANTAS, José Ricardo Fonseca[11]

SILVA, Lorranne Gomes da[12]

Introduction

The practice of skateboarding is now globalised, as well as being a sport whose health benefits for the body and mind are undeniable. It is therefore essential to understand the importance of this practical activity in the lives of its practitioners, in order to promote appreciation, support and respect.

Skateboarding developed in the United States, mainly in the state of California, dating back to the 1920s. Inspired by traditional pushcarts, people improvised what would later become the skateboard. It didn't take long for it to arrive in Brazil, initially in Rio de Janeiro in 1965. In the 1970s, skateboarding gained momentum and polarised mainly on the Rio-São Paulo axis.

In the city of Quirinópolis, the practice of skateboarding is recent and dates back to the 1990s, with the first skateboards brought from Ituiutaba and Uberaba. During this period, skateboarding was less intense and there were no appropriate places for it to take place, only on the streets and pavements. From the 2000s onwards, skate rinks were created, which contributed to safer and more specialised places to perform manoeuvres, train and have fun.

The aim of the research was to highlight the practice, spatialisation and profile of skateboarders in the city of Quirinópolis/GO in 2015, as well as the importance of urban space and local public policies for the sport to intensify, helping more people and acquiring new fans.

Authors such as Araújo (2010), Armbrust (2008, 2010), Brandão (2007, 2008, 2010, 2012), Carlos (2007), Machado (2015), Santos (2000), among others, were

---

10 This article is the result of monographic research defended and approved at the State University of Goiás, Quirinópolis Campus, in November 2015, as part of the Geography course.

11 Graduated in Geography at the State University of Goiás, Quirinópolis Campus. .kokefonseca@hotmail.com

12 Professor of Geography at the State University of Goiás, Cora Coralina Campus. PhD in Geography from the Federal University of Goiás, Institute of Socio-Environmental Studies, Goiânia Campus. .lorrannegomes@gmail.com

fundamental to the writing.

1 - The origins of skateboarding

The first reports of boards on wheels date back to 1880 (GALLO, 2012). Therefore, according to data from the Brazilian Skateboard Confederation (2014), in 1918, an American boy had already dismantled skate axles and wheels and fixed them to wood, but he didn't ride standing up, but with one knee resting on the wood and the other foot giving impetus.

Skateboarding first developed in the United States, especially in the state of California, around the 1920s, and then spread to other parts of the world. According to studies by Machado (2008a), inspired by traditional pushcarts, people improvised what would later become skateboards. In 1959, skateboards, which had previously been handmade, began to be manufactured by major brands and sold on a small scale.

Camargo (2012) states that with the drought and the scarcity of waves in California (USA) in the 1960s, surfers appropriated skateboarding, giving it a new meaning as a sport. Regarding the first skateboarding practices, data from the Brazilian Skateboarding Confederation (2014) shows that:

> At first it was called pavement surfing, and it quickly spread throughout the United States. By 1963, sidewalk surfing, already practised by a large number of teenagers, had created an identity, with its own manoeuvres, and thus gained its definitive name: Skateboarding. The first skateboarding championship took place in Hermosa Beach, California, in 1963, won by Larry Stevenson. This was followed by the launch of the world's first specialised skateboard magazine, The Quarterly Skateboarder, in 1964 in the United States.

However, at that time the materials used to make skateboards were not of the best quality, resulting in a series of injuries to skateboarders (CAMARGO, 2012). In the early 1970s, with the invention of polyurethane wheels[13] , *skateboarding* became more common, and the first championships and the first big names in *skateboarding* emerged (MANOSA, 2009). From then on, it slowly became disconnected from **surfing** (MACHADO, 2008a).

Skateboarding therefore represents a form of human cultural expression in the process of creating, innovating and re-adapting new elements. This means that skateboarding is capable of appropriating the space-time process to materialise and leverage new horizons.

---

[13] Plastic material with good abrasion resistance.

The origin of skateboarding encompasses formats that space makes up with socio-urban influences. Society thus becomes responsible for the spatialities that are gradually and temporarily formed. Deursen (2009) also confirms that the city has always been the setting for this sport and has been responsible for the transformation of skateboarding over time. Thus, we realise that, based on culture, expressions, the imposition of man on skateboarding and skateboarding on man, it becomes the responsibility of skateboarders to lead countless innovations in this new urban art based on human creationism.

However, more than art, the main focus of skateboarding to be portrayed is urban mobility, which is being applied in a new world of mutual possibilities, where skateboarders interact and modify all the apparatuses imposed by the physical urban system.

In this context, the history of skateboarding has been marked by adaptations from the rolling cart, where the wheels and metal spheres interacted to move the object, to modern wheeled objects, at which point creativity began to be introduced based on the adaptations that practitioners needed.

According to Araújo (2010), the first skateboarding formats were Downhill, Pool Riding and Freestyle. Downhill was an adaptation of those who wanted the sensation of surfing on tarmac, going downhill. Soon after, during the great drought in California, Pool Riding emerged, a different project where empty swimming pools were used for skateboarding, the so-called **bowl** .[14]

Freestyle, on the other hand, is the discipline that features manoeuvres in sequence, usually on the ground, and gave rise to most skateboarding manoeuvres. For Bastos; Reckziegel and Stigger (2005, p. 2), a fourth manoeuvre module to be described is Street, which emerged

> Due to the closure of several rinks in the USA and favourable urbanisation, skateboarders began to ride on the streets using pavements, stairs, walls or anywhere they could ride. With the assimilation of Freestyle (not widely practised) and vertical skateboarding, Street skateboarding developed, as the name implies. Later, Street Skateboarding began to be practised in specific places, such as squares or parks, places that skateboarders call rinks. These places have obstacles, ramps, stairs and handrails, etc? They imitate common obstacles in street furniture.

But it's worth considering that, within these manoeuvring modalities, there are the characteristics produced by the different body techniques that influence the aesthetics of gestures and movements that contain marks of the plurality existing in skateboarders'

[14] A type of skate park in the shape of an empty oval swimming pool used for pool riding.

lifestyles. According to Bandão (2012, p. 17):

> Seen as a mix of leisure and adventure in an era marked by greater social control, these activities offered their practitioners an alternative to expressing excitement in public - albeit in a moderate way - and a certain antidote to the tensions resulting from the continuous effort of self-control and social restrictions.

Thus, we can see that the remarkable trajectory of skateboarding clarifies and highlights various actions and elements that have provided new meanings for "skateboarding" which, in a way, have been transformations in the improvement of the object, expanded by human action, interspersing the spatialisation of cultural movements.

The period during which these events took place, together with the need to expand creativity and reinvent new skateboarding activities, formed a very strong and innovative cultural quantity, which came to be established in a concrete and lasting way. In this context, Brazil appears as an ally that has restructured the entire link between competitiveness and innovation in the practice and the materials produced, as the following text shows.

2 - Skateboarding in Brazil

Skateboarding didn't take long to arrive in Brazil and one of the factors that mediated the emergence of this sport in the country was globalisation. According to studies by Chaves and Britto (2000), skateboarding arrived in Brazil in Rio de Janeiro in 1965, brought by Americans.

At first, skateboarding in Brazil, according to Brandão (2011), was taken up by the bourgeoisie, and only in 1970 did it take hold and expand to other regions. In 1977, Machado (2008b) describes the appearance of the first skateboarding magazine in Brazil, **Esqueite**, a translation into Portuguese, with only two editions that presented the scenario of this new sport. In 1978, another magazine appeared, **Brasil Skate**, with three editions aimed at publicising the recent practice of skateboarding and presenting manoeuvres, the construction of ramps, the talents of the time, among other things.

One element that occurred in relation to the first publications of content about skateboarding in magazines in Brazil was that, in the 1970s, there was a striking link, because Brazilian skateboarders, in particular, were influenced by North American skateboarding (MACHADO, 2012). They saw that national magazines were trying to reproduce North American skateboarding, but with the particularities of their own Brazilian culture (MACHADO, 2008a).

The 1970s also saw the inauguration of Latin America's first skateboarding rink in Rio de Janeiro, revolutionising Brazilian skateboarding with the construction of two **bowls**, which led to a change from the free style practised on the streets to the Pool Ridding style. This rink hosted Brazil's first rink championship. Its architecture was influenced by **Skateboarder** magazine, which showed Americans in backyard rinks and pools performing manoeuvres (HONORATO, 2013).

Although skateboarders could be found in several Brazilian states during the 1970s, Rio de Janeiro emerged as the main skateboarder on the national scene, due to the strong connection between surfing and skateboarding when this activity was first introduced in Brazil. However, in the 1980s, the city of São Paulo began to claim the title of "skate city", according to Bastos (2006). This happened at the end of the 1970s, when skateboarding began to decline, and at the beginning of the 1980s, according to Dias (2011, p.35):

> Skateboarding survived thanks to the few remnants of the sport who built private ramps and organised competitions. Even in this phase of decline, the sport relied on the construction of new places to practice and the organisation of some events. The structures built were the Paço Municipal and Wave Cat tracks, in São Bernardo do Campo, and the Bowl at the Itaguará Country Club, in Guaratinguetá (SP), in 1981.

One of the characteristics of the 1980s in Brazil, especially during the process of opening up and re-democratisation that took place after 21 years of military rule, was the strengthening of youth protagonism. This could also be seen in the skateboarding movement, as Fortes and Brandão (2013) point out. Regarding this period, Brandão (2012) considers that the political libertarian act of the Brazilian nation boosted the young people who skateboarded. So, rather than simply travelling around the city, they began to take it as a place of interpretation, reading it in the most diverse ways.

At this time, according to Dias (2011), several specialised TV programmes were launched, as well as the creation of more specialised magazines, such as **Overall** (1985), **Yeah!** (1986) and **Skatin** (1988/1990). In the artistic movement, according to Souza (2008), punk and rock bands such as Legião Urbana, Garotos Podres, Titãs, Camisa de Vênus, Plebe Rude, Nenhum de Nós, Replicantes and Biquíni Cavadão began to emerge.

Regarding the influence of music on the practice of skateboarding in Brazil, Fortes and Brandão (2013) consider that it is possible that, among the sports that shaped subcultures in Brazil in the 1980s, skateboarding was the one in which music was most important, with the musical taste shared by the members being a relevant issue within the subculture. This led to the emergence of an undeniable link between punk rock and skateboarding.

Brandão (2010) adds that skateboarding's association with the punk movement soon resulted in skateboarding being completely removed from the realm of surfing and brought it more into contact with the streets and the stylised rebellion of youth social movements than ever before. For Fortes and Brandão (2013, p. 219):

> In this context, attitude became the crucial element of a skateboarder. Skateboarders moved around the city, horrifying the masses and streamlining the space. People were amazed by skateboarders, both because of their appearance, the manoeuvres they performed and their use of equipment and urban space.

Bastos' (2006) studies show that during the early 1990s there were indications of stagnation. Skateboarding went from a very large universe, as it was the "explosion" of the time, to then falling sharply, leaving few fans of the sport. The disappearance of skateboarding, according to Dias (2011), was so considerable that it was almost impossible to buy a new skateboard and the most assiduous practitioners made their own objects from skateboard parts and pieces of wood.Thus, in the course of the 21st century, Brandão (2011) considers that skateboarding was the target of media exposure such as it had rarely experienced in its recent history. High-rated television programmes began to interview skateboarders or show them performing "radical manoeuvres", highlighting this cultural practice. According to Bento (2013, p. 1):

> It's interesting to note that, despite the assimilation of the activity by the market and the media, skateboarders continue to face a certain amount of prejudice in society. A prejudiced association with the "lifestyle" of skateboarders persists in the popular imagination, even though there are as many lifestyles as there are practitioners.

In 2016, skateboarding became a sport that spread throughout Brazil and the world, gaining more and more fans. The following text will show how skateboarding is practised in the city of Quirinópolis/GO.

3 - Practice, spatialisation and profile of skateboarders in the city of Quirinópolis/GO

3.1 - Research methodology

The research was qualitative and the composition of the data presented was based on the methodological procedure of participant observation which, according to Goldenberg (1997, p. 63), "is a research methodology that appears as a possibility in the attempt to describe, explain and understand the object of study", also used to compose the data from bibliographical surveys and interviews.

The interviews were open-ended, a model which, according to Boni Quaresma (2005), allows the interviewees more freedom and autonomy to discuss the topic suggested by the interviewer.

Among the questions that made up the research were: what were the places where skateboarding took place in Quirinópolis from 2000 onwards; what were the skateboarding spots in the city until 2015; how are these spaces historically and culturally constituted in the imagination of the people who experience this practice? What are the possibilities for keeping alive this cultural heritage located in the urban space of Quirinópolis?

Thirty skateboarders were interviewed, all of them male and aged between 11 and 27. The age range was very different and the highest incidence of skaters was between 15 and 22 years old, which clarifies the fact that skateboarding is not only linked to young people, but also to adults. These individuals are necessarily those who have lived in Quirinópolis for more than 10 years and have a linearity in this sport, practising it whenever possible and bringing new things to skateboarding through competitions and championships.

Of those interviewed, four have been skateboarding for more than 10 years, seven are skateboarders who have been doing it for between five and nine years, while the rest have been doing it for less than five years. This favours a strengthening of the intensive continuation of skateboarding in the city.

It should be noted that within the skateboarding disciplines there are the following practices: Downhill[15] , Pool Riding[16] , Freestyle[17] and Street[18] **. Street skateboarding** is one of the fastest growing sports in terms of the number of fans in Brazil, which is why we decided to investigate this sport in Quirinópolis.

3.2 - Results and discussions

The municipality of Quirinópolis is located in the south-west of Goiás, a region that stands out in the state's economy due to investments in agro-industry. The city has approximately 47,950 inhabitants according to the IBGE (2016). Since 2008, with

15 Skateboarding practised on steep slopes.
16 Skateboarding in empty oval pools.
17 Skateboarding on flat, smooth terrain.
18 Skateboarding is practised in every urban space.

numerous national and international investments, it has developed into an important centre for the production of biofuels, especially ethanol, subsidising two large-scale sugar and ethanol plants.

Due to its flat urban space, typical of the south-western region of Goiás, the city is favourable for skateboarding, whose origins date back to the 1990s, brought from Ituiutaba and Uberaba in Minas Gerais. The influence of these municipalities on Quirinópolis is probably due to the geographical proximity of these three cities.

The first skateboards brought to Quirinópolis came from these two cities in the state of Minas Gerais, which soon led to the search for skateboard parts and materials from these places. From the 2000s onwards, the city of Rio Verde/GO also began to play an important role in the commercialisation of skateboarding products, because at that time the logistics of Quirinópolis did not fully meet the demand for skateboarding parts and equipment, taking into account its minority of fans at the time.

In 2012, a **skate shop** was set up in the city of Quirinópolis in order to meet demand and take advantage of the potential that the city was providing with the consumption of parts and materials for **street** skateboarding. As a result, skateboarding in Quirinópolis has become stronger and more recognised in the logistics sector to this day.

During the initial investigations carried out in 2014 and 2015, ten main skateboarding spots were located in Quirinópolis, four of them as specific places for everyday skateboarding and the other six classified as sub-points[19]. These are considered secondary due to the oscillation in the frequency of skateboarders at these locations, at times considered high and at others low.

This classification stems from the different logic that exists in these locations, which are scattered around the city and are known by skateboarders as **gaps**, or street skate peaks.

The four specific or main points are: the Municipal Covered Market; the obstacle course in the Antônio Roberto Chaves Woods; the Sol Poente Lake skate park and the mini **ramp** (Children's Square), and the sub-points are: the staircase of the Municipal Teotônio Vilela Theatre; the Workers' Terminal (Anésio Caiado Square); the Municipal wake hall; Private space 1; Private space 2 and Private space 3, spatialised in figure 01:

[19] Uvinha (2001) explains that **when skateboarders used "routes" to get to the rinks,** they usually passed through points (subpoints or secondary peaks), because their movement through the urban space took place in such a way that the streets favoured a dynamic interaction with various points during their journeys, thus creating a range of skateboarding activities.

The following text will briefly describe each point.

Point 01: Woods

The Antônio Roberto Chaves Forest was built in the Conjunto Rio Preto neighbourhood in Quirinópolis in the 1980s and revitalised by the city's mayor in 2012, so that the city's residents could have a wooded leisure area where they could see the exotic animals that were locked up in cages. There was also a park for children to encourage walking and running.

During this period, seven sports courts were built: two for futsal, two for shuttlecocks, one for bocce, one for volleyball and one for tennis to meet the social, cultural and sporting needs of Quirinopolina society. The tennis court, which wasn't used much because there weren't many practitioners of the sport between 2004 and 2006, became the skateboard court of the Antônio Roberto Chaves Woods (Figure 2).

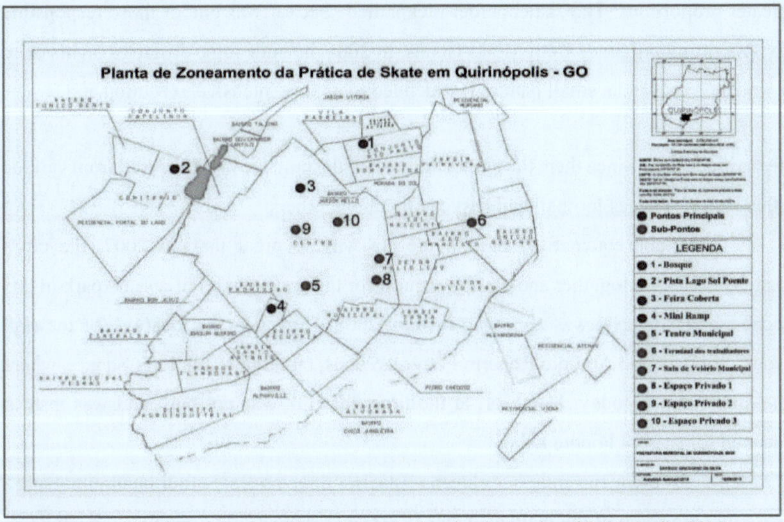

Figure 01: Spatialisation of skateboarding points and sub-points in the city of Quirinópolis, 2015. Source: Urban zoning plan for the municipality of Quirinópolis/GO Organisation: DANTAS, José Ricardo Fonseca.

Figure 2: Quirinópolis/GO. Skateboard court in the Antônio Roberto Chaves Woods
Source: DANTAS, José Ricardo Fonseca / Jan. 2010

In 2006, with influences brought from Uberlândia/MG, this place began to be appropriated by skateboarders and, from then on, skateboarding in Quirinópolis took on greater proportions. The skateboarder nicknamed "Secão" was one of those responsible for this strong influence from Minas Gerais. In 2006, he built some obstacles on his own, a wooden crate and a small handrail, which led to a more intensive execution and

diversified sport. Since then, the number of fans of the practice has increased from various other sports, such as football and basketball, among others.

The consequence of this simple act was so great that, in 2007, the city's skateboarders got together and made a request for the construction of a skate park in this location. The request was approved in May 2008. However, the location for the rink would not be in the Antônio Roberto Chaves Woods, but in the Children's Square, where the **mini ramp** is today. However, at the time, the rink was not built and was instead planned for the Sol Poente Lake.

The Grove currently has eleven obstacles: three crates[20] , four handrails[21] solo - these two are some of the main elements of **street** skateboarding -, a delta[22] , a caraia[23] , a minifubox[24] and a polyjam[25] . The degree of difficulty of these obstacles is the same, as they are basic in skateboarding; what differentiates them is the difficulty of the manoeuvre performed on each one.

The Bosque court is large and can be used by several skaters at the same time.

[20] Wooden obstacle with a metal corner.
[21] Round metal obstacle with ground support.
[22] Two ramps facing each other.
[23] Straight ramps from 30° to 45° are used for climbing, manoeuvring and descending again.
[24] Obstacle with ramp, handrail and crate together.
[25] Metal pipe fixed to the ground at an inclination of 40°.

The Bosque is open from eight in the morning until ten at night, all week long, and it also has four reflectors to help skaters practise at night. On average, thirty skateboarders go there every month.

Point 02: Lago Sol Poente track

The Sol Poente Lake skate park (Figure 3), inaugurated in October 2008, was **designed** by Quirinópolis skateboarders themselves, between the Capelinha and Tonico Bento neighbourhoods.

The construction of the track had some modifications to the initial project, in terms of the quality of the material and the angles and degrees, causing problems in the use of some obstacles, as one of the interviewees mentions:

> Some of the obstacles became much bigger and different from what had been proposed, and another important factor was the quality of the materials used to build the track, because when it (the track) was built, the floor cracked and had to be replaced, and today the floor is 45 per cent damaged, with several cracks, without any maintenance (INTERVIEWEE A).

Figure 3: Quirinópolis/GO. Lago Sol Poente skate park
Source: DANTAS, José Ricardo Fonseca / Mar. 2015

Another problem is the lack of drinking fountains, since this is a public place for physical activity, there is a shortage of water to hydrate the body, favouring social well-being. In 2012, Lago Sol Poente hosted its 1st Regional Skateboarding Championship, with skaters from Rio Verde, Santa Helena, Goiânia and Uberlândia taking part. The track has several obstacles, but skateboarding is only practised during periods when there is less sunlight, due to the lack of trees around the track.

The obstacles on the skate park at Lago Sol Poente are large. Some of them are approximately 3 metres high. On site, there are: two 45°[26], a spinne[27], a trojan horse[28], a pyramid[29], two handrails, a crate and four quarters[30]. These obstacles allow local skateboarders to reach high levels of difficulty in more elaborate manoeuvres. Through observation, it was possible to establish that, on average, this track receives fifteen skaters a month.

Point 03: Covered Market

According to the skateboarders interviewed, the Municipal Covered Fair (Figure 4), located in the Central Sector, was one of the first places where skateboarding was practised more intensively, around the 2000s. In the beginning, there were few skateboarders and practically no tracks or obstacles for manoeuvres. From 2003 onwards, skateboarding began to be practised at this location, thanks to the purchase of some wooden obstacles, the cost of which was sponsored by the city council at the time.

Figure 4: Quirinópolis/GO. Municipal covered market.
Source: DANTAS, José Ricardo Fonseca / Nov. 2014

Used for skateboarding when it rains and other places are unsuitable for skateboarding, the covered market has a well-proportioned area with a smooth floor, steps and is suitable for skateboarding during the week and on Saturdays[31]. The busiest times are in the afternoon. These obstacles represent a degree of difficulty that ranges from the simplest, using the floor, to the highest and most dangerous, using the stairs, which are very long and high.

The covered fairground is also used for skateboarding events. The first was in

---

[26]A straight 45° ramp is used for climbing, manoeuvring and descending again.
[27]Obstacles with two ramps facing each other, which can be straight ramps or quarters.
[28] A small 45° accompanying a small quarter.
[29]Obstacle that has a table and at least three sides.
30Half of a Half or mini-ramp.
[31]It was noted that after the market on Sunday and Monday, the amount of rubbish is enormous, which gets in the way of skateboarding.

2003, when some wooden obstacles were inaugurated. Another event that stood out was Skate Rock, which took place in 2013. This event was organised by the Quirinópolis skateboarders themselves, with the help and funding of local businesses.

Point 04: Mini Ramp

The Children's Square was built in the 1980s by the mayor of the Promissão neighbourhood to serve the population. The square only had a children's playground and a few snack bars. When it was revitalised in 2011, the park was refurbished, some coconut trees were replanted, an outdoor gym and a shuttlecock court were built. In addition, a mini-ramp was built in the Children's Square (Figure 5) in 2010, further strengthening the practice of skateboarding in Quirinópolis.

Figure 5: Quirinópolis/GO. Mini ramp. Children's square
Source: DANTAS, José Ricardo Fonseca / Mar. 2015

It is known that the mini-ramp was built with inferior quality and inadequate labour, leading to cracks and unfinished obstacles, which has resulted in its very low use by skateboarders. The only obstacle located in the Children's Square is the mini-ramp itself, and it is up to the practitioner to perform countless manoeuvres with varying degrees of danger on its concrete structure.

Skateboarders would usually gather in the evenings to perform manoeuvres on this ramp. Today, however, this track is hardly frequented and no championships are held there. In the speech below, it is possible to understand the situation experienced by skateboarders from Quirinopolino today, in relation to the infrastructure present in the city:

> Skateboarding itself needs a structure for the sport to survive. Unfortunately, every infrastructure has to be built at a price and has to have a specific location. And Quirinópolis, due to its political culture, still doesn't meet the conditions of the municipality's social status, the release of funds from the state, because it only happens after a tender for public works and skateboarding is so

> important to the practitioner that he takes the conditions out of his own pocket for the construction of rinks, which means yet another adapted place to practice (Interviewee B).

It can be seen in this statement that public policies in Quirinópolis do not meet the structural needs and suitability that young skateboarders need in order to practise. With regard to the factor of organisation and commitment to duties that the city's politicians and administrators have, it can be seen that they don't take their role of ensuring order and laws seriously, as laid down in the Brazilian federal constitution.

Analysing the frequency of use of street skateboarding locations by skateboarders, Graph 1 was drawn up, showing the distribution of practitioners at the four main points mentioned in Quirinópolis:

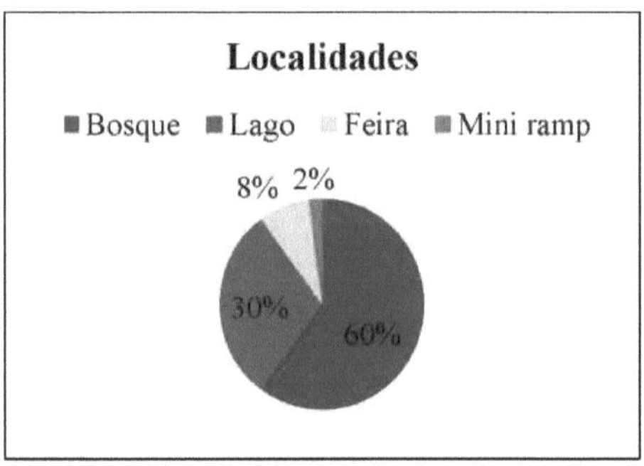

Graph 1: Quirinópolis/GO. Main skateboarding locations. 2014, 2015
Source: DANTAS, José Ricardo Fonseca

As shown in graph 1, the Antônio Roberto Chaves Forest is the place most frequented by skateboarders. This is probably because it is a favourable environment for street skateboarding[32] , with a smooth court and lots of trees around the court, which provides a pleasant climate for skateboarding.

Sub-point 05: Municipal Theatre

The Municipal Theatre (Figure 6) is located in the city's Central Sector, a sub-location intended for the use of this physical urban space for skateboarding in Quirinópolis.

There are reports that the practice of skateboarding at this location dates back to 2003, with the staircases, benches and the open-air stage being used as obstacles for street

[32] Street skateboarding.

skateboarding . One of the interviewees highlights his motivation for skateboarding at this location, showing that street skateboarding in Quirinópolis has good potential when it comes to varying the location: Interviewee C said: "The choice to skate at the theatre is because it has the stairs and the stage, which are very good for street skateboarding, because skateboarding only on the lake is quite monotonous, not only here, but in many other places in the city."

Figure 6. Quirinópolis/GO. Staircase of the Teotônio Vilela Municipal Theatre.
Source: DANTAS, José Ricardo Fonseca / May 2014

In this speech, it was noted that the skateboarders' connection with urban spaces is related to their actions within the structure of the city of Quirinópolis. The interviewee described how they don't just use one spot in the city to skate, but several other spots, which they find very satisfying and enjoyable.

Sub-point 06: Workers' Terminal

The Workers' Terminal strengthens the spread of skateboarding in the city. There's a logic of companionship and social gathering of skateboarders who establish their relationships, ideals and practices involved in the world of street skateboarding.

The site was inaugurated in January 2008, where the Anésio Caiado Square was established, located in the Onício Resende neighbourhood. The structure built on this site has a lot of potential for skateboarding and provides several benches and pavements, which makes for great street skateboarding action in the city (Figure 7).

Interviewee D explains that skateboarding requires specific physical structures, which are damaged over time and, as a result, skateboarders adhere to certain elements, such as paraffin to move the structure and some other repair materials, as interviewee D pointed out: "I like to do a *street* in the Workers' Terminal, because, bro, the edges that you have on the benches are really good to walk on, you just have to apply a lot of candle or in some places that are damaged, you just have to apply plastic putty and it's beautiful."

Figure 7. Quirinópolis/GO. Workers' Terminal. Anésio Caiado Square.
Source: DANTAS, José Ricardo Fonseca / Sept. 2015

It is understood that, for skateboarders from Kirinopolino, there is a strong adaptation exercised on the urban architecture of this square when they talk about using some objects to be applied to the concrete structures. In this case, as stated in the initial municipal engineering project, the square's benches would only serve as seating. For skateboarders, as well as being able to sit down, it's also possible to use the square's benches as a good obstacle for **street** skateboarding.

Sub-point 07: Municipal viewing room

The Sala de Velório, located in the Centro neighbourhood, is also on the list of skateboarding hotspots in Quirinópolis. The reason skateboarders chose this spot was because of the side access ramp that is located there. Skateboarders, however, use this structure differently from other people, using the ramp as an obstacle. The report of the use of this location begins in 2007, with the spread and use up to the present day, as can be seen in Figure 8.

Figure 8. Quirinópolis/GO. Funeral parlour.
Source: DANTAS, José Ricardo Fonseca / Aug. 2015

Interviewee E said:

> The only problem is the tarmac, which is becoming potholed, making for a bad landing, as it catches the wheels, favouring several tumbles that can hurt.

According to the skateboarders, there have never been any reprisals for banning skateboarding here. What's more, as they have specific behavioural ethics of a human nature, they respect the place and are also respected.

Sub-point 08: Private space 01

Private space 01 (Figure 9), located in the Centro neighbourhood, is most often the place where skateboarders gather to chat, skate and go to other parts of the city.

This place has always been welcoming, because the skateboarder who brought the influences of skateboarding from Uberlândia settled there and has always been willing to develop and help any elements that are linked to skateboarding in Quirinópolis, such as: donating used skateboard parts to people who wanted to ride, donating trainers to people who couldn't afford them; always with the aim of supporting skateboarding in the city. As a result, skateboarding only takes place on the ramps on the pavement, on the little handrail and on the street, where they perform basic ground manoeuvres, but they are quite technical. One of the interviewees confirms the importance of the location for practising the sport:

> Look, this place, since I started skateboarding, everyone hangs out at Secão, even the kids who come from other cities to ride and compete when there are skateboarding championships in the city, they also stay at this place, because this is where the masses hang out, they all get together to skate and ride (INTERVIEWEE F).

Figure 9. Quirinópolis/GO. Private space 01.
Source: DANTAS, José Ricardo Fonseca / Oct. 2010

In interviewee F's speech, it is clear that the place is of particular importance in skateboarding in Quirinopolis, as it is characterised as a set of links and information that are organised and regrouped when they say that they come together to strengthen the ideas and practices of the spatialising process in terms of sport and leisure in **street** skateboarding.

Sub-point 09: Private Space 02

According to skateboarders in Quirinópolis, one of the private spaces (Figure 10) described in this research is the ideal place for those learning to ride GAPS - street obstacles - because the height of the staircase in its structure is favourable for performing manoeuvres.

Figure 10. Quirinópolis/GO. Private space 02.
Source: DANTAS, José Ricardo Fonseca / Dec. 2014

According to them, this location is good for all levels of skateboarders, as the staircase faces the street, allowing more space to perform manoeuvres that are not only simple, but also complex. For interviewee G: "Dude, skateboarding on this staircase is classic, because it's low and small, you can do a lot of manoeuvres, I really like skateboarding here and it was on this staircase that I lost my fear of jumping off high obstacles and that was really good for me."

This location is very interesting, so that at weekends or during the night, the incidence of movement at this peak becomes much lower, with practically no movement that could get in the way of skateboarding, as Interviewee H points out:

> The problem is that you can only go at night and on weekends. Regarding the ban on skateboarding in this place, there are no reports, but there is the recommendation that the guard asks people not to skate near the glass door and asks us to skate more on the side of the stairs towards the street (INTERVIEWEE H).

It's clear from interviewee H's comments that this is an important point for skateboarding in Quirinopolino, especially for those who are street skateboarders.

Sub-point 10: Private Space 03

Another private space (Figure 11) mentioned in this monograph is a place in the city's Central Sector, where there is a high potential for street skateboarding, because it also has a structure that is very rich in elements that favour a good performance when it comes to this practice.

Figure 11. Quirinópolis/GO. Private space 03.
Source: DANTAS, José Ricardo Fonseca / Mar. 2015

Interviewee I points out in the following speech that the architecture of the place is suitable for skateboarding:

> This place is good for me, it has a wall with a large space of smooth floor, and that's perfect, old man. Not to mention the fact that you can skateboard even on the street, you know it's quite difficult, but you can try and manoeuvre on the floor of the place. (INTERVIEWEE I).

According to the skaters, this spot is used by them in the mornings and afternoons, and is generally more frequent at weekends, but there are some skaters who hang out during the week, usually those who don't work and have more free time. There is also a shop there, but it works the other way round, as it is a snack bar that is only open at night.

Nothing was discussed about the place having prohibitions, as they reported that they never asked the owner if they could skate or not, they just said that they do and that they don't care what people say, since the place is open, without any restrictions so far. They also say that they skate during the period when the establishment is closed and take the opportunity to walk around and practise their sport or leisure.

The sub-points make it possible to widen the field of practice for the obstacles, covering an entire urban perimeter and raising the technical level of the manoeuvres, as they are very close to the streets and pavements in Quirinópolis. In this way, the athlete or practitioner finds themselves in the midst of society during practice, with the presence of cars and people travelling near these locations.

The choice of a particular secondary peak is linked to the physical structure of the location, which can be of sufficient quality for skateboarding. For the most part, these locations include the concrete of buildings, such as steps, staircases, stages, handrails, smooth floors, ledges, benches, etc., which become very specific obstacles for **street** skateboarding. One of the interviewees agrees and goes on to say:

> [...] the media constantly influences each practitioner, but because skateboarding is an individual sport that degrades the aesthetics of buildings, it gives a bad social impression, not to mention the aggressive style of each practitioner, where they expose their personality to the maximum. (INTERVIEWEE J).

This causes a lot of damage and friction with the obstacles, thus causing the destruction of some places and making the skater's image somewhat obstructed. However, as will be described in the next topic, it can be seen that skateboarders from Quirinopolinos are privileged with a common sense of friendship rather than conflict, seeking, for the most part, just the possibility of having fun.

Final considerations

It was noted that the practice of skateboarding in Quirinópolis is recent and dates back to the 1990s, with the first skateboards brought from Ituiutaba and Uberaba. During this period, the practice was less intensive and the appropriate places were just the streets and pavements. From the 2000s onwards, the spaces for skateboarding in the city became more varied, with the use of specific tracks with obstacles to perform manoeuvres and, consequently, more fans of the sport or leisure.

Skateboarders are generally young people who are looking for another way to use the urban space of Quirinópolis, with the aim of having fun through a sport that contributes to the well-being of the body and mind. Thus, the urban space perceived and experienced by skateboarders takes on other conceptions and values.

The skateboarder, as an inhabitant of the city, transforms it, recreates it, with his form of art and culture. This clearly demonstrates the potential that the city can offer them. In this context, it was noted that skateboarders can give the shape they want to any object or obstacle in the urban space, expanding the possibilities of their practice and their intense creativity.

As well as contributing to physical health, skateboarding also contributes to people's mental health, providing moments of leisure and tranquillity. It is therefore important that there is more investment and public policy to make skateboarding more accessible in the city of Quirinópolis.

However, despite being a worldwide sport, unfortunately the way skateboarders are treated still reveals prejudiced behaviour that needs to be reflected on socially.

References

ARAÚJO, A. G. *Landshapers:* Freestyle territory, a geography of skateboarding. In: 1st Colloquium on Autonomous Territories. UFRJ, 2010.

BASTOS, B. G. The lifestyle and social trajectories of skateboarders: from "neighbourhood" to "running". Dissertation (Master's in Human Movement Sciences), 174 f, 2006.

BASTOS, B. G.; RECKZIEGEL, A. C. C.; STIGGER, M. P. Skateboarding as a culture to be studied: ethnography as an option. In: Proceedings of the XIV Brazilian Congress of Sports Sciences, XIV CONBRACE. Porto Alegre, 2005.

BENTO, M. Skateboarding: a brief overview of the practice in Brazil, Argentina and Chile. Contemporâneos: Revista de Artes e Humanidades. n.11, nov./abr. 2013.

BRANDÃO, L. Publicidades da rebeldia: Sport and consumption in the rhythm of **punk rock**. Contemporâneos: Revista de Artes e Humanidades. n.5, nov./abr. 2010.

_________. The City and the Skater Tribe: Youth, Daily Life and Bodily Practices in Cultural History. Dourados: UFGD, 160p, 2011.

_________. From the transfigured city to the transformed city: Youth cultures and the practice of skateboarding (1970/1980). History and Culture Magazine. Franca, SP, v.1, n.2, p.7-20, 2012.

BONI, V. & Quaresma, S. J. (2005). Learning to interview: how to conduct interviews in Social Sciences. Revista Eletrônica dos Pós-Graduação em Sociologia Política da UFSC, 2(1), 68-80. Available: <www.emtese.ufsc.br>. Accessed: 20/09/2016.

CHAVES, C.; BRITTO, E. A onda dura: 3 décadas de skate no Brasil. São Paulo: Parada

Inglesa, 114 p, 2000.

CAMARGO, I. B. Skateboarding in Porto Alegre in the 1970s: A history of the first generation. Monograph (Bachelor's Degree in Physical Education), 69 f, 2012.

CBSK (2014). History of skateboarding in the world. Brazilian Skateboarding Confederation. 2014. Available at: <www.cbsk.com.br>. Accessed: 06/03/2015.

DEURSEN, T. W. V. About skateboarding in the city. Ciência do Skate E-zine, year 1, issue 1, nov. 2009.

DIAS, G. O. Skateboarding beyond sport: social manifestation and cultural movement. Dissertation (Master's in Sociology), 190f, 2011.

FORTES, R.; BRANDÃO, L. Anarchic, punk, "unlabelled": skateboarding in Fluir and Yeah! magazines. Communication, Media and Consumption. São Paulo, ano 10, v.10, n.27, p.211-236, mar. 2013.

GOLDENBERG Mirian. The art of research: how to do qualitative research in the social sciences. Rio de Janeiro: Recorde, 1997.

HONORATO, T. The sportivisation of skateboarding (1960-1990): Relations between the macro and the micro. Rev. Bras. Ciênc. Esporte, Florianópolis, v.35, n.1, p.95-112, jan./mar. 2013.

IBGE. Brazilian Institute of Geography and Statistics (2016). Available at: <cidades.ibge.gov.br>. Accessed on: 20/09/2016.

MACHADO, Giancarlo (2008a). Skateboarding and Globalisation. Available at: <www.youtube.com>. Accessed: 06/03/2015.

_________. Historical record of Brazilian skateboarding. Skatecultura (2012). Available at: <www.skatecultura.com>. Accessed on: 01/06/2014.

MANOSA, R. S. Organisational analysis of Element Skateboards. First module final paper (Marketing Technologist), 131 f, 2009.

CHAPTER 4

PRAÇA CÉLIO MORAES ANDRADE IN VILA PROMISSÃO: a place of diversity and contradiction in ways of living

Onilda Pereira de Jesus[33]
Marlane Aparecida Cabral[34]

Introduction

The square is still a privileged space for citizens, even in the face of the various existing urban configurations and despite the structural, temporal and functional transformations of the public space. And this is not only because it brings people together, but also because it enables various manifestations, due to its multifunctional character, which people appropriate and where they relate to each other (CALDEIRA, 2007). The square stands out in terms of the formation and development of Brazilian urban memory, because the uses and appropriations of these spaces reflect the customs of a society.

If we consider the square as belonging to the **lu gar** category in the geographical classification, and as a collective space, we can say that it is the synthesis of complex urban life. Yes, the public square is the soul of the city, as it comprises a multiplicity of dynamic and complex relationships.

Thus, considering the urban space of Quirinópolis, the Célio Moraes Andrade Square was singled out for an analysis of the relationships manifested in its daily life and its surroundings, in other words, the relationships between the people who live in the vicinity of the square mediated by this space and the relationships nurtured by these residents.

The aim was to verify the ways in which this public space is used, as well as its redefinition and the ways in which people relate to it, following its revitalisation in 2012 and, above all, after a free market was set up there. Finally, the intention was to demonstrate the social representativeness of Praça Célio Moraes Andrade and also to draw people's attention to the social, historical and cultural value of Quirinópolis' public squares.

The novelty of this research makes it relevant, given that there are no in-depth studies on Célio Moraes Andrade Square and its surroundings and on the social relations

---

33 Graduated in Geography from the State University of Goiás, Quirinópolis Campus. .onildapj@yahoo.com.br
34 Professor of Geography at the State University of Goiás, Quirinópolis Campus. .marlaneac@hotmail.com

that run through it, as well as their respective meanings. Furthermore, there is little research in Quirinópolis that deals with the square as a symbol of social identity, a point of articulation and an essential place in the city's daily life; nor does it deal with the functions, uses and appropriations of this collective space.

To organise the study, one of the methodological procedures adopted was bibliographical research. The methodology also consisted of field visits to observe the movement of people and analyse social interaction in the square, with a view to applying questionnaires and interviews with local residents and visitors. Documentary research was also carried out using documents archived at Quirinópolis City Hall (bills, requests and newsletters), photographs, maps and tables.

The importance of place in the construction of geographical knowledge

Discussing the concept of **place** within geographical science is essential, as it is a fundamental means of understanding human reality in the contemporary world. For Lisboa (2007), place means much more than just a geographical location, it relates to the various types of experience and involvement of the subject with the world.

According to Santos (1988), it is worth revisiting the debate about whether Geography is a science of places, as Vidal de La Blache pointed out, or whether it is a science of men. From his point of view, however, even if the way man inserts himself into the place is linked to the history of the present, Geography tends to gradually be "the science of places created or reformed to fulfil certain functions" (SANTOS, 1988, p. 52). These different functions are attributed to the same place by history. According to Professor Holzer (1999), Lukermann was one of the first to reflect on the locational properties of place and to define Geography as the science of places. Lukermann apud Holzer (1999, p.69) says that:

> The culmination of geographical study is the description of the Earth in geographical order. The key to this order lies in the locational concept of place. Emphasising the relative, the cultural, the historical experience of humanity, in relation to the physical attributes of the area, is to make the complete study of geography - the study of places.

According to Holzer, humanist geographers, based on phenomenology, have been concerned with defining place as an experience, which essentially concerns space as it is experienced by human beings. In other words, place would be a core "generator of geographical meanings" (HOLZER 1999, p. 70), which is dialectically related to what is called "space". In this sense, Holzer mentions that Tuan, in prioritising the study of place from the perspective of localisation, proposed that:

> [...] the place is a unit among other units linked by the circulation network; [...] the place, however, has more substance than the word localisation suggests: it is a unique entity, a 'special' whole, which has history and meaning. Place embodies people's experiences and aspirations. Place is not just a fact to be explained in the broad structure of space, it is reality to be clarified and understood from the perspective of the people who give it meaning (TUAN, apud HOLZER, 1999, p.70).

For Bartoly (2011), the contribution of the humanist current of Geography to the in-depth study of the concept of place is undeniable, however, he realises that the critical current and some more recent approaches open up different perspectives for advancing the analysis of the definition, saying that for Critical Geography:

> [The specificity of the place comes from the role it plays in the capitalist dynamic. In defining the concept, it is not necessary for a place to enjoy longevity for it to be relevant, but rather for it to fulfil an important function that sets it apart from the rest. As a result, a place stands out and therefore becomes unique, based on its greater or lesser technical and communication capacity, for example. As soon as a place has an adequate functional structure, one gets the impression that the world necessarily passes through the place, that global flows not only pass through it, but establish a dialectical relationship with it. (BARTOLY 2011, p. 68)

Thus, for the critical current of Geography, place is no longer understood only as lived space, with a phenomenological slant, but tends to be considered a social construction. Thus, for the more critical thinkers, according to Bartoly (2011), it is the relationships between global flows and local conditions that give rise to place; in other words, place is defined through the relationships it maintains with the totality driven by the historical movement of capitalism. The particularity of place, then, is conditioned by historical time, given that the meaning of place changes as capitalism changes.

According to Bartoly (2011), the tension between the two geographical visions gives rise to a place made up of a material dimension and an abstract dimension. The first would be "the one that tells us about the location of objects, their objective distribution in space and the relationship of the place with the totality" and the second dimension, in turn, would be "the one that deals with the symbols and meanings that are attributed by individuals to the place and give meaning to the very arrangement of objects and people in this portion of geographical space" (BARTOLY 2011, p. 68).

Carlos' (2007, p.17) propositions conform, in a way, to the two currents of thought in Geography. For her, place is where "life develops in all its dimensions", since it is the product of human relations. That's why her analysis (of place) is crossed by the idea of a construction, woven by social relations carried out on the plane of the lived (everyday life), which "guarantees the constitution of a network of meanings and senses that are woven by the history and culture of civilisation that produces man's identity" (CARLOS, 2007, p. 22).

According to Carlos (2007, p. 22), place is the point of articulation between the world in constitution and the concrete place, where "imbalances, conflict situations and trends in society" are expressed in relation to the world. Chaveiro (apud MARTINS, 2015, p.55) also points to a conceptualisation of place that is in line with Carlos', when he states that:

> [...] place is the social, historical, cultural and symbolic movement that takes place - and is invested - in everyday life [...]. It is therefore the task of place to involve the weaves of the mode of production, technical objects, cultural manifestations, subjectivity, the actions of subjects and, therefore, the involvement of different corporealities in the use and appropriation of space.

In this context, it is clear that favouring just one of the geographical currents can lead to a reduction in the importance of place as a substantial component of the individual's identity, or in other words, to approaching place only as an abstract phenomenon, devoid of materiality and objectivity.

The space of everyday life: a place where local and global relations are experienced

The relationships that exist between individuals and the place make it possible for identities to emerge and strengthen. Even if economic, social and cultural changes have occurred, individuals and/or societies will always retain an identity with the place. As Carlos (2007) points out, these spaces can be felt, thought about, appropriated and experienced through the body; it is the space of everyday life where these relationships manifest and are determined.

In this context, it is the neighbourhood, for example Vila Promissão, that brings us back to the immediate space of life of the closest daily relationships experienced by the residents, such as neighbourly relationships like "going shopping, walking, meeting acquaintances, playing ball, games" (Carlos 2007, p.18).

At first, these banal, everyday and familiar actions seem meaningless, but according to Carlos (2007), they are what create deep bonds of identity. For Morais (2001), it is precisely in these trivial aspects, through the experience and practices established by the residents, based on recognising the other and the sense of belonging to a certain place and group, that the relationships established between the individual and the place gain relevance.

For this reason, it is necessary to understand the everyday functionality that is actually exercised between the places of use that are constituted in space (MARTINS, 2015). And what is everyday life, in terms of its characteristics and meanings? In Lefebvre (apud Gil and Gil Filho):

> Everyday life is a sum of insignificance, not insignificance. Banality is important in life, getting up, opening the window, enjoying the weather, feeling the sounds and smells of the day at dawn. It's the banality of everyday life that makes our lives meaningful in this world. We must also consider that there are different degrees of banality. The author reads everyday life from the point of view of instrumental rationality, in other words, for him, capitalism has colonised everyday life, creating an everyday life full of meanings and signs, where consumption is directed and encouraged by capitalist production, which provides the basis for the constitution of capitalism as a system for reproducing social relations of production. Thus, banal everyday life becomes the object of capitalist social organisation (LEFEBVRE apud GIL and GIL FILHO, 2008, p.106)

Damiani (2001) and Carlos (2007) see everyday life as a set of affections, activities and relationships, a fabric made up of ways of being and ways of living that are specific to each resident, which take place in the place and its temporalities, generating a plurality of meanings. In agreement with Lefebvre's thinking, the authors state that the rhythm of everyday life is marked by the strategies for reproducing the social relations of the capitalist mode of production. Carlos (2007, p.54 and 55) emphasises that:

> (everyday life) today is impoverished in the sense that it is increasingly subordinated to the world of commodities, in which signs of status permeate relationships, and exchange value subjugates and captures the meaning of use. From this perspective, everyday life is presented as the place of repetitive gestures and the uniformity and homogeneity of habits, forms of use, behaviour, values, etc. All programmed by capitalism and the state strategy that organises everyday life (how people live, dress, how and where they spend their leisure time) in an attempt to abolish difference.

This is because "everyday life is a level of analysis of the totality", since through "the mediation of everyday life in place, we are taken from particular facts to global society" (DAMIANI 2001, p.162 and 164). The author also states that "to think of everyday life in place is to think that the world is in place and place in the world". (DAMIANI 2001, p. 168).

Place, according to Carlos (2007), is thus the point of articulation between the world in constitution and the concrete local, where "imbalances, situations of conflict and the tendencies of society" are expressed in relation to the world. As "each subject is situated in a space, place allows us to think about living, dwelling, work and leisure as lived situations, revealing the conflicts of the modern world on a daily level". (CARLOS 2007, p. 22 and 20).

Relph also emphasises the importance of man's relationship with place, saying that "a deep relationship with places is as necessary, and perhaps as inevitable, as a close relationship with people; without such relationships, human existence, though possible, is devoid of much of its meaning". (RELPH apud HOLZER 1999, p.72).

Looking at Célio Moraes Andrade Square and its surroundings from this point of view, we can recognise the affective ties of the neighbourhood's older residents in terms of their identification and sense of belonging to the place. In addition, the square articulates neighbourly relations between adults and children, who use it as a space for leisure and entertainment (games, informal conversations, etc). After being revitalised in 2012, Praça Célio Moraes Andrade has also become an environment for sporting activities, including walking.

For some time now, commercial and economic relations - also crossed by labour relations - have permeated the daily lives of the residents around Praça Célio Moraes Andrade, due to the Agricultural Exhibition that takes place near the square, the activities of the Rural Union and the trade in various products in establishments such as a bar (brothel), a grocery shop, an accessories shop, sandwich shops, a car garage and a Fire Brigade unit. Now, the square itself and the streets around it are used as commercial outlets for the open market that takes place every Friday evening.

Therefore, we can understand place through what is manifested in everyday life, both through relationships that occur at a local level, where relationships of identity and belonging are built, linked to subjectivity, habits and behaviour, and through human relationships that interact with broader contexts, interconnected with the reality of other groups - economic, political - and which are not dissociated from global systems.

Identity and subjectivity of place

Place is the result of human relationships, of social relations woven by socialising, and which has meaning and significance produced by the history and culture of a people. It is in this sense that many authors point to place as a reality that can be perceived and analysed through the individual's personal experience.

It is possible to conceive of space without the interference of human beings, but not of place. As Barreto (2013) sees it, just like man, place requires an identity, a name. According to the author, just as people have a body, places have a kind of corporeal extension. So, from the perspective of subjectivity, for Barreto, wanting to be with someone is not very different from wanting to be somewhere.

Subjectivity, identity and a sense of belonging are categories that make up a place (SILVA, L. apud SILVA, M. 2014, p. 16), produced from living with other people, from intersubjective intentions and experiences and also through the way we adapt to and appropriate the global realities that are inserted into the local and the recognition of localisation references (BARTOLY 2011).

Subjectivity is the interpretations, perceptions, emotions, thoughts and feelings of each human being that derive from their own life experiences, beliefs and values. When it comes to place, subjectivity is the dimension of living, of experience, because as Tuan apud Holzer (1999, p.74) says, experience constitutes places on various scales. According to Holzer (1999, p.74), experience forms a continuum that encompasses "the home, as the primary provider of meanings; the city, as the centre of meanings par excellence; neighbourhoods and regions; the nation state".

The feeling of belonging to a part of space, as Bartoly (2011, p.72) also states, can be related to the street, the neighbourhood, the state or the country. "When I'm in my neighbourhood, my spatial identification is my street". And when we walk through the city and see the name of the neighbourhood where we live on a sign, we quickly recognise that name as part of us.

> **The neighbourhood thus appears as the place where a** social **'engagement' manifests** itself, or, in other words: an art of living with partners (neighbours, shopkeepers) who are linked to you by the concrete but essential fact of proximity and repetition (MAYOL apud MORAIS 2001, p.108).

That's why Carlos (2007, p.17) states that the place, such as the neighbourhood, the square "is the portion of space that can be appropriated for life [...]" because it is appropriated by the body, the senses and even the footsteps of its inhabitants. In other words, the subject belongs to the place just as the place belongs to the subject, so that "inhabited places - marked by presence, created by fragmentary history [...], nature transformed by social practice, the product of a creative capacity, cultural accumulation that is inscribed in a space and time" (CARLOS 2007; p. 22) are interconnected with the social nature of identity and the feeling of belonging to a place. In short, for the author, identity, which is complex and refers simultaneously to the local and the global, is generated in place, the representative and consolidator of social relations and practices.

Hespanhol and Moreira (2007, p. 51) state that place only acquires identity and meaning through the intentions that human beings attribute to it, since it "is the product of lived personal experience, permeated with symbolic, cultural, political and social dimensions". According to Martins (2015, p. 28), the subject is also the being who represents the place or gives it meaning, which is manifested through "symbols and everyday relationships that mould and (re)create manners, customs and culture" [...].

In this way, identity is formed through unconscious processes, but it is not innate, natural, nor definitive and immutable, in fact it is incomplete "[...] it has components and forms that vary with changing circumstances and attitudes" (RELPH apud FERREIRA 2000, p.68), so it remains in formation and constitution over time. Relph understood that the process of developing a place's identity consisted of:

> A combination of observation, i.e. direct contact with the place, and expectations established prior to this contact. The identity of a place would thus be the expression of the adaptation, assimilation, accommodation and socialisation of knowledge. (RELPH apud FERREIRA 2000, p.68)

Place would then be an indispensable axis of meaning for the foundation of a human being's identity as an individual and as a member of a community, thus coming to be associated with the concept of home. According to (TUAN apud HOLZER 1999), the intersubjective intentions and experiences - the fruit of familiarity - that make up the process of identity construction begin at birth and deepen with experience, resulting in a detailed knowledge of place and the constitution of roots.

For Buttimer apud Hespanhol and Moreira (2007, p. 52), as for Relph, even if there are transformations in the place, both for the individual and the community, there is the feeling that the old characteristics remain, reinforcing the identity with that place. Considering that identity is formed by static physical location, activities, meanings and also by the spirit of the place, according to (FERREIRA apud HESPANHOL e MOREIRA, 2007, p. 52), "the deeper one is inside a place, the stronger the identity with it".

One example is the Célio Moraes Andrade Square and the Vila Promissão neighbourhood: although time has passed and there have been changes to the physical structure and landscape, as well as social and economic changes, some residents remember the past and feel a sense of belonging to the place.

However, as Massey apud Ferreira (2000, p. 75) states, place cannot be thought of only from the "perspective of unique identities, constructed through deep and historical relationships", since place would not have a single meaning, shared by all people. The identity of places is better understood if placed in the plural, since places have varied identities and are full of relationships with the world.

Rethinking the identity of place today therefore requires us to consider that it is increasingly dependent and constructed on a global level, meaning that "the history of the place increasingly passes through the shared history that is produced beyond the physical limits of the place, that is, its specific situation" (CARLOS 2007, p. 21).

If the relationships developed in the place are not isolated from the world, nor detached from the capitalist system and the globalisation process, it is necessary to reflect on the identity of places, bearing in mind that:

> **The proliferation of pasteurised spaces, based on "artificial values"** such as consumption, and the very constancy and speed with which people move, would promote a drastic reduction in the intensity of an **"authentic"** relationship **with the place (BARTOLY 2011, p. 75).**

In this sense, contemporary life would present us with a world in which rootedness, relationships of identity and belonging are becoming fewer and fewer. Consequently, the production of the humanist place would be threatened.

Quirinópolis: the place and its history

The municipality of Quirinópolis belongs to the south-west region of the state of Goiás and is part of the geographical micro-region of Quirinópolis (made up of Lagoa Santa, Itajá, Itarumã, Caçu, Cachoeira Alta, Paranaiguara, São Simão and Gouvelândia) and the Meso-region of Sul Goiano. Its geographical position is determined by the coordinates 18° 26'52" South latitude and 50° 27'07" West longitude and its territorial area is 3,786.694km^2. The estimated population is 47,377 inhabitants, with a population density of 11.41 inhabitants per square kilometre (IBGE, 2014).

Farming was the mainstay of Quirinópolis' economy for many years, due to its flat, fertile land. From the 1960s onwards, agricultural and commercial market activities linked to rice and bean production developed. With the growth and modernisation of agribusiness in south-west Goiás in the 1970s, the cultivation of monocultures such as corn, soya and cotton began to emerge and expand in Quirinópolis (SANTOS, G., 2012).

During this period, Quirinópolis' traditional cattle farming also grew, to the point where it became one of the largest municipalities in the state of Goiás in this economic activity. Until 2005, Quirinópolis was an agricultural and grain producer. From then on, two sugar and alcohol plants from the state of São Paulo, São Francisco and Boa Vista, set up in the municipality.

Due to the implementation of sugarcane plantations, other agricultural activities began to "divide up the productive spaces, with sugarcane taking over the predominance of cultivable areas from 2008 onwards" (SANTOS, G., 2012, p.33). The expansion of sugar cane plantations has strengthened the city's role as a producer of sugar, ethanol, electricity and biodiesel for the national and international markets.

Although the implementation of sugar-alcohol activities in Quirinópolis has led to the dispersal of a considerable number of small rural producers through leases and sales of rural properties, their presence is still very strong (SANTOS, G., 2012). These small producers are responsible for the diversified production of products (fruit, vegetables, eggs and meat) in the municipality, destined for consumption by the producing families and for sale in the local trade, at open-air markets such as the one that takes place in Praça Célio Moraes Andrade every Friday.

The origins of the municipality of Quirinópolis date back to the mid-19th century, with the arrival of several farmers from Minas Gerais and São Paulo in the region, which until then had been completely unpopulated. The motivation for the arrival of the explorers was the 10-year exemption "from paying taxes on cattle and horse breeding for farmers who settled in the south-west of Goiás" (D'ABADIA 2010, p. 45).

> The city of Quirinópolis was one of several attempts to colonise the Brazilian hinterland, back in the 1830s. Its construction is a good example of a historical moment when the colonial frontiers advanced towards western Brazil (CORRÊA E XAVIER 2012, p. 185).

The first inhabitants, then, were farmers like Captain João Chrisóstemo de Oliveira e Castro, who arrived in the region with his family and slaves around 1932. In 1840, it was the turn of Mr José Ferreira de Jesus, his family and slaves to arrive in the future settlement. Two years later, Custódio Ferreira do Prado and his family also settled on these lands. In addition to these, the Rodrigues family and José Ferreira de Jesus were also pioneers in the region (MATTOS and PARREIRA 2010).

José Ferreira de Jesus and his wife Maria Jacinta de Oliveira donated a piece of land to the Church in 1843, so that after the donation, the priest from the Rio Verde parish began to visit the site and perform Catholic religious rites. Later, in 1868, under the leadership of José Ferreira de Jesus, a chapel was built approximately two kilometres from the Rio das Pedras, between two tributary streams. One of these streams had no name and was called Córrego da Capela (Chapel Stream) because of the small church. Thus, a settlement was founded, which was called Capelinha (MATTOS and PARREIRA 2010).

The town's first business was Casa Comercial A. Rodrigues e Cia, "a haberdashery and fabric shop and a large warehouse" (D'ABADIA 2010, p.46), owned by Colonel Antônio Rodrigues. It was the Chapel and the Commercial House that enabled the settlement and growth of the region and, consequently, the emergence of the municipality. On 29 July 1879, the town was elevated to a collective parish with the name of Nossa Senhora D'Abadia do Paranaíba (MATTOS and PARREIRA 2010). On 10 July 1894, the parish became a district of Rio Verde, but kept the same name.

The district of Nossa Senhora dabadia do Paranaíba was at the foot of the Serra da confusão do Rio Preto and was therefore a damp and unhealthy place, prone to disease. These inadequacies prompted the town to move and be founded on a new site. In 1913, the foundation stone was laid three kilometres away from the old settlement and, in 1919, the new town was inaugurated (D'ABADIA 2010). On 31 December 1943, the district was elevated to the category of municipality, with the name Quirinópolis, in honour of Colonel José Quirino Cardoso. On 22 January 1944, Quirinópolis was officially emancipated from Rio Verde.

Place as a perceived and affective space: Célio Moraes Andrade Square in the Vila Promissão neighbourhood

When considering the concept of place, the study of Praça Célio Moraes Andrade analysed the relationships that occur between the residents of the Vila Promissão neighbourhood and the square on a daily basis, through the accounts of these individuals, since their experiences, memories, habits and culture are what make up this place and make it complex and symbolic. As the place is understood on the basis of the practices developed within a historical dimension, it was decided to take a look back at the formation of the municipality of Quirinópolis, as well as its economic aspects, in addition to the history of the Vila Promissão neighbourhood. It was also necessary to analyse the particular history of Célio Moraes Andrade Square: its construction, inauguration, revitalisation and the changes it has undergone. In addition, we investigated what has been built, organised and structured around the square, influencing and altering this place as "a consequence of the process of constituting the world" (CARLOS, 2007, p.17).

In the city of Quirinópolis, as in the majority of municipalities in Goiás, growth took place in the form of neighbourhoods[35] and villages[36] . However, the urban fabric of

[35] A neighbourhood is a community or region within a city or municipality, and is the minimum unit of urbanisation in most cities around the world.

[36] Vila is a population agglomeration of intermediate size between village and town, with an almost self-sufficient economy, which characterises its transition from village to town, in which

these areas did not always follow the concept that gave them their name. Thus, the names of neighbourhoods are often preceded by "Vila", as is the case with the Vila Promissão neighbourhood, even though the word has another meaning (CORRÊA E XAVIER, 2012).

The Vila Promissão neighbourhood, where the Célio Moraes Andrade Square is located, arose "from the construction, with resources from the State and Federal spheres, of housing estates financed for the population by COHAB (Goiás State Housing Company)". The neighbourhood was created in the 1970s and was characterised by being a kind of geometric framework, "a set of narrow streets, as well as their pavements, which was probably designed this way so that it wouldn't escape the urban layout of Santos Dumont Avenue and Rio Preto Street" (CORRÊA E XAVIER, 2012, p.188). According to Sagim and Sagim (2000), Praça Célio Moraes Andrade was built by Quirinópolis city hall during the second term of Onício Resende's administration (1988-1992) and was called Praça do Sindicato because it was located in front of the municipality's Rural Union building. Célio Moraes Andrade Square is located at the intersection of Vitória, Francisco Corrêa Neves and Alameda Paranaíba streets, as shown in map 1:

In March 1996, councillor Célio Rosa do Prado asked the Quirinópolis City Council to name the square Célio Moraes Andrade, in honour of the son-in-law and right-hand man of former mayor Onício Resende. During Onício's second term, Célio Moraes Andrade held various positions in Quirinópolis City Hall and, after the end of the administration in 1992, Célio Moraes Andrade went on to work as a manager for Agropecuária Onício Resende Ltda. On 21 December 1994, Onício and Célio died in a plane crash. Thus, in April 1996, by means of Law No. 2,114/96 of the Quirinópolis City Council, Praça do Sindicato was renamed Célio Moraes Andrade. This law also authorised the construction of a bust of the honoured man in the square, a project that has not yet been carried out.

---

the tertiary sector (commerce and services) is of significant importance and provides it with some economic self-sufficiency.

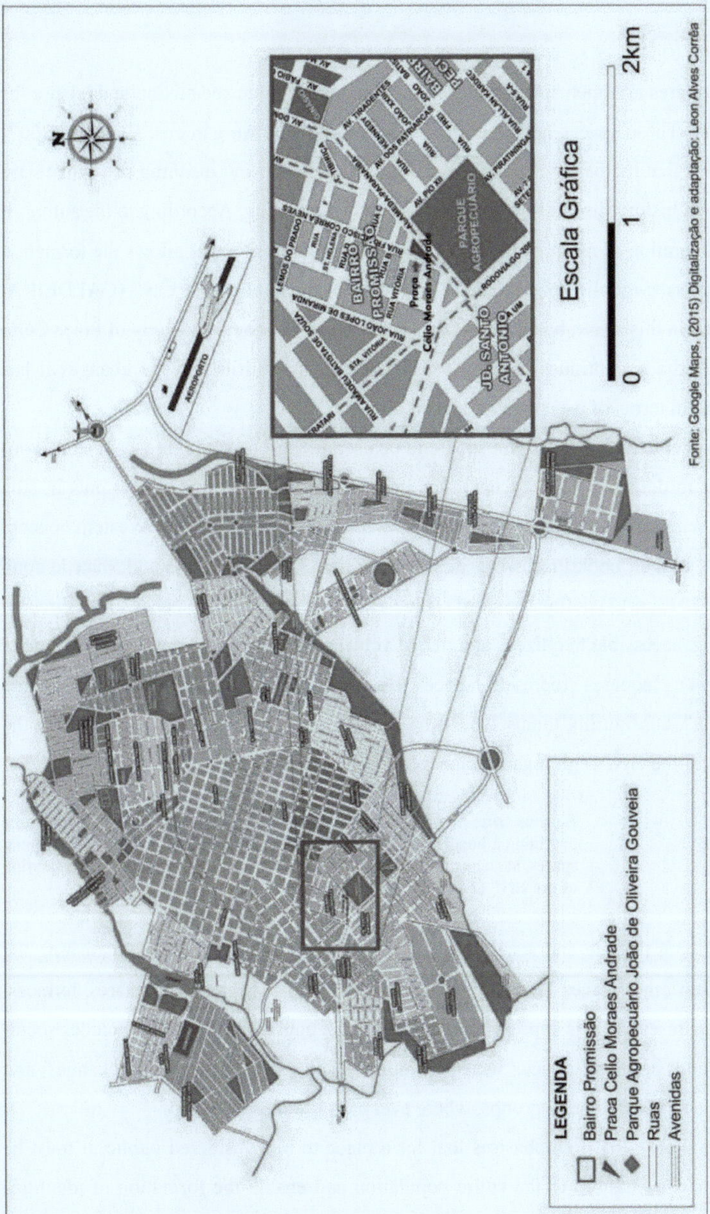

Map 1: Location of Célio Moraes Andrade Square
Source: Google Maps, 2015. Digitisation and adaptation: León Alves Corrêa

The square as a public place: a collective space of appropriation, use and multiple subjects.

Squares are open spaces whose function is to provide socialising and leisure for their users. These spaces are generally associated with urban gardens. Caldeira (2007) emphasises that the beauty, social and cultural importance and meaning of a square are based on its history, its landscape design and its urban setting. According to the author, it is "the integration of morphology, aesthetics and appropriation that allows the formation of squares as symbolic spaces, places of memory, the soul of the city" (CALDEIRA, 2007, p.3). In this sense, based on the notion of collective space, the study of Praça Célio Moraes Andrade continues to focus on the functions it fulfils and the changes it has undergone in terms of use and appropriation.

It is common for the term square to be used in various contexts today, so it is not so simple to define it as a concept, even because, as Segawa (1996, p.31) puts it, "the square is an ancestral space that is confused with the very origin of the Western concept of urban". In their book Praças Brasileiras (Brazilian Squares), Robba and Macedo apud Paglioto (2015, p.13) define the word square as "urban public open spaces for leisure and socialising, accessible to citizens and free of vehicles". As an urban element, for Caldeira (2007, p.4), "squares represent spaces of sociability favourable to meeting and socialising", configuring themselves as true nodes of social confluence, essential to the daily life of the city. With regard to the morphology of the city:

> Squares differ from other spaces in that they represent gaps in the urban fabric, providing a break in the landscape formed by buildings. They are referential spaces, acting **as visual landmarks and "as focal points in the organisation of the city"** (ZUCKER apud CALDEIRA, 2007, p.13).

Taking the use and accessibility of space as a principle, but distinguishing the square from other places in the urban fabric, such as parks, gardens, squares, terraces, etc., in this research, it is approached as a collective, public space, which includes socio-cultural and political issues, which favours meetings, recognition, exchanges and coexistence between social groups, where everyday life is reproduced.

Paglioto (2015) emphasises that for a place to be considered public, it must be symbolically accessible to the entire population and enable the formation of identities both among those who frequent it and between the inhabitants and the place. Therefore, to designate a square as a public place "is to recognise it as a space where the diversity of subjects and uses gives it a symbolic character capable of marking collective identities and cultural references" (PAGLIOTO 2015, p. 36). Therefore, it is this multiplicity of uses (ways of using) and subjects (people who use) that give the square its status as a public space.

In view of this, as corroborated by the discussion made earlier, about place being a product of human relations, a social construction (embodied in history, politics and economics) produced and producing intersubjective experiences, which generates plural meanings and manifests itself in everyday life, we can maintain that the square is a symbol of a social identity, since:

> The symbols that make up a social identity are not totally arbitrary or random constructions, since they always maintain certain **links with concrete reality [...] it seems consensual that "identities"** are always constructed from the recognition of an alterity. However, this can only happen where there is interaction, transactions, relations or contacts between different groups (SERPA apud PAGLIOTO 2015, p.21).

Magnoli also contributes to the understanding of the square as a representative and transmitter of the social relations built and culturally organised in the city and subject to the process of identification and the affirmation of collective identities, when he reinforces that in it "the multiplicity of society, its dramas, the confrontation between citizenship rights come to light. Different social experiences are condensed, conflicts, perceptions and ambiguities are exacerbated: the whole is reiterated in its parts" (MAGNOLI apud PAGLIOTO, 2015, p.14).

The research carried out on Célio Moraes Andrade Square exemplifies this, as it identified various forms of use, for example for parking, pedestrian crossings, as well as uses for formal and informal commerce. In each activity and practice carried out in the square, multiple subjects were perceived, who create and nurture relationships between themselves and with the place, relationships that make the place a generator of meanings, a moulder and disseminator of identities.

Analysing the characteristics of the square's physical structure and the species of trees that were planted there before the renovation (Figure 1), by comparing old and current photos, we found that Célio Moraes Andrade Square was planned to function as a green area, as an extension of the "green" in the city, providing shade and purifying the air. After the revitalisation, however, the square lost this function because the trees were cut down.

Figure 1: Célio Moraes Andrade Square before revitalisation
Source: CANDIDO, Elias (2009)

Observing the configuration of the square (the external pavement in the shape of triangles forms parking spaces), it was concluded that it was planned to serve mainly as a car park, used by clients and employees of the Rural Union, users of neighbouring buildings and the public of the João Oliveira Gouveia Agricultural Park during the Agricultural Exhibition and other events on site. As such, the Célio Moraes Andrade Square's function as a car park is not the result of a new type of appropriation, as occurs in other squares during changes of use; it was preconceived for this purpose. This way of using the square remained even after its revitalisation.

Another function of the Célio Moraes Andrade Square is that of a pedestrian crossing, and as it is the point of convergence of two busy access roads to other neighbourhoods and to GO 206 (Rua Francisco Corrêa Neves and Alameda Paranaíba, respectively), the sense of passage extends beyond the daily circulation of people in the Vila Promissão neighbourhood.

Caldeira (2007, p.188) states that in modern times, and especially in contemporary times, the square has been established "as an appendage to other functions, such as circulation and the road system", since the role of movement in the city has become a fundamental element of the urban configuration.

> The square takes on the role of a place of passage, a carousel, a junction. Its traditional function as a meeting place, the vital centre of urban life and the structuring node of the city, is replaced by alternatives that encourage the production of empty spaces (CALDEIRA, 2007, p.188).

According to Robba and Macedo, analysed by Paglioto (2015, p. 36), in the course of the 20th century, the attributions given to squares underwent changes, which allowed the resumption of some activities that were common in public spaces in the colonial period of Brazil and that had been abandoned. Commercial and service use (TABLE 1) is an example of "uses that were added to the square's programme, as a legacy of the colonial tradition of squares".

TABLE 1: Evolution of the squares' programme of activities

FUNCTION OF THE SQUARE	PERIOD			
	COLONIAL	ECLECTIC	MODERN	CONTEMPORARY
	Socialising Social	Contemplation	Contemplation	Contemplation
	Religious use	Tour	Recreation	Recreation
	Military use	Scenario	Sports leisure	Sports leisure
	Trade and fairs	Socialising	Socialising	Socialising
	Circulation		Scenario	Scenario
	Recreation		Cultural leisure	Cultural leisure Commerce Services Pedestrian circulation

Source: ROBBA; MACEDO apud MASSAE (2012).
Organisation: JESUS, Onilda Pereira de. (2015)

Célio Moraes Andrade Square has been used for commercial activities for almost 30 years, when a snack bar - located at one end of the square - was given to the first owner by Quirinópolis City Hall, during the second term of Onício Rezende's administration (19881992). According to the current owner, Wellington Vieira Lopes, the shop has had many owners and repairs to its structure over the years.

The place is currently a sandwich shop called 'Lanche Chapa Quente', which is open every day of the week in the evening and attracts not only residents from around the square and the Vila Promissão neighbourhood, but also from other parts of the city. In this respect, the sandwich shop also provides socialising in the square through its customers.

On the opposite side of the square, there is another trading post: a stall owned by 79-year-old Mr Arnor José dos Santos, who sells vegetables, flour, preserves, home remedies, among other things to supplement his income as a pensioner. The stall is open during the day from Monday to Saturday. To be closer to the square, he moved to the Vila Promissão neighbourhood about six months ago. For Arnor, the stall is more than just a source of profit, it's also a place for interaction, because as well as customers, many people stop by to chat to him. In addition, some employees from Usina São Francisco "exchange ideas" near the stall while they wait for the bus to go to work.

The retired man has been working there for three years and, during the exhibition period , he rents the place out as a snack bar. However, the commercial use of the square that attracts the most people, has reconfigured daily life and changed relations in the area is the open-air market, which took place in Praça Célio Moraes Andrade after its revitalisation.

It's worth noting that Célio Moraes Andrade Square doesn't fit the romanticised vision of a square - a picturesque space for religious festivities - that is often ingrained in the collective imagination. In Quirinópolis, Praça Coronel Jacinto Honório, also known as Praça da Velha Matriz, is the one that most closely matches this profile, since it was planned with these characteristics when it was renovated in the 1960s[37] , as Neves (2011, p.52) recalls:

> [...] The Praça da Matriz, with its luminous fountain (coloured waters dancing to classical music) where the inhabitants could have fun and improve their knowledge of the more elite musical culture, sanitary facilities, a new bandstand and gardens, reserving a place for the **installation of a water mirror, where ducks and geese were raised. [...] In the** square, the Church celebrated the Feast of the Patron Saint, when the sacred and the profane came together. At the same time there were masses in honour of Our Lady of the Abbey and a large number of stalls set up by the faithful who attended the festival and a variety of shops catering for needs such as eating, dressing, entertainment and the purchase of household utensils.

This doesn't mean that Célio Moraes Andrade Square wasn't used as a space for meetings, leisure or socialising - it was and still is. As has been said, being open and free, squares create conditions for sociability, they are places of aggregation, of confluence, since this character is inherent to them.

Over time, the social representativeness, degree of importance and cultural significance of squares for the community have fluctuated due to the growth of urban violence, the lack of maintenance of public spaces, changes in the dynamics of how cities work, people's lifestyles and values, and what is meant by leisure and leisure places (PAGLIOTO 2015).

But all is not lost, because each historical moment is characterised by new ways of using spaces, which take on new roles. In this sense, Andrade and Bovo (2010, p. 3) criticise the deep pessimism that is often attributed to squares in a generalised way today. According to the authors, it is not possible to "translate the function, use and meaning of squares, since the essential characteristic of such places is the cultural diversity of possibilities they can harbour and not one specific function or another". Caldeira (2007) also argues that the functionalities of the public square are not fixed and rigid:

> The square represents a kind of chameleon-like space, capable of changing and adapting to the transformations of cities, allowing for diverse appropriations. This peculiarity has meant that the square has historically acquired a diversity of forms and functions, without losing its essence as a collective space (CALDEIRA, 2007, p.14).

[37] The new facilities at Praça Velha Matriz were officially handed over in August 1968.

From this perspective, it can be understood that over the years and especially after the revitalisation, the functions and uses of Célio Moraes Andrade Square were redefined (Figure 2). The old trees were cut down and small palms and other small trees were planted in their place; the paved area was restructured, the grass area was modified, the commercial points were repaired and new benches were installed. Before, the square didn't have any toilets; after the renovation, two chemical toilets were put in place. After the revitalisation, a bus stop was installed in the square, which is now used by rural workers from Quirinópolis' sugar and alcohol mills as a public transport stop.

Figure 2: Célio Moraes Andrade Square after revitalisation

Source: CANDIDO, Elias (2012)

Thus, it was observed that the site performed different functions before and after its renovation, so that some types of use remained even after the revitalisation. Not only time, but human actions on the place modify and re-signify it. Furthermore, as the square is an essentially plural space capable of adapting to the transformations of the urban space, it acquires and allows for a variety of forms, functions and appropriations, without losing its main characteristic, which is that it is a collective space.

Final considerations

When studying Célio Moraes Andrade Square, we found that it is a place capable of constructing collective and individual identities, it is a symbol of the social identity of the Vila Promissão neighbourhood, where the relationships expressed and constructed are linked to the forms of use and appropriation by multiple subjects. The use and appropriation addressed here is not the simple act of using the square, but, as Carlos (2007) argues, it is what marks the relationships between people and the space on a daily level.

Direct affective relationships were identified, such as those of belonging, coexistence and leisure associated with the use of the square as an extension of the home, as a place to meet, interact with neighbours and family, as a distraction, pastime and space for children to play. Direct social relationships were also perceived, which are not always based on emotional ties, for example: those of functionality, associated with the use of the square as a car park, as a pedestrian crossing and as a structuring/configuring element of the urban fabric.

Indirect social relations were also found in the research, linked to the use of the bus stop and commercial/consumer relations, through the appropriation of public space for the free market, the 'Lanche Chapa Quente' sandwich shop and the stall of Arnor José dos Santos. The research also found that Praça Célio Moraes Andrade acts as a mediator of indirect social relations, with the agricultural exhibition in Parque João Oliveira Gouveia and the Rekints Bar (brothel).

It was noted that there is a psychological contradiction among the residents of the area around the square regarding the organisation of the "cattle festival". The festival is eagerly awaited every year because it is the biggest cultural event of its kind in the city, providing entertainment through rodeo competitions and great artistic shows, as well as helping to generate income for the municipality. On the other hand, some of these same residents complained about rubbish in the streets, invasion of privacy due to vehicles parked on the pavements of their homes, noise pollution and disturbed sleep.

There are also contradictions in the position of some local residents/frequenters regarding the impressions and sensations generated in the square. At the same time as beneficial feelings of tranquillity, safety and freedom were attributed to the place, there were mentions that the activities of the brothel contribute to increased violence and marginalisation in the locality. The establishment was also characterised as prone to inappropriate scenes, a generator of insecurity, detrimental to tranquillity and unsuitable for families living in the neighbourhood.

Finally, without wishing to exhaust the discussions raised, we have tried to record that the square is an excellent sample of the city from a historical, cultural and urban spatial point of view, where it is possible to identify the representation, construction and transmission of multiple social relationships. In this way, the diversity and contradictions in ways of living can be seen.

References

ANDRADE, Thiago Bocon; BOVO, Marcos Clair. The meaning, use and social representativeness of urban public spaces: the case of São José and Getúlio Vargas squares in Campo Mourão/ PR. Paper presented at the 5th Scientific Production Meeting, Campo Mourão, PR, 2010. Available at: <http://www.fecilcam.br/nupem/anais_v_epct/PDF/ciencias_exatas/01_ANDRADE_BOVO.pdf> Accessed on: 26 August 2015.

BARRETO, Paulo Irineu. On the humanity and subjectivity of "place". Portal Página Cultural. Uberlândia, 2013. Available at:< http://paginacultural.com.br/sobre-a-humanidade-e-a-subjetividade-do-lugar/>. Accessed on: 21 August 2013.

BARTOLY, Flávio. Debates and perspectives on place in geography. Geographia, Rio de Janeiro; v. 13, n° 26, p. 66-91, 2011. Available at: <http://www.uff.br/geographia/ojs/index.php/geographia/article/viewFile/454/325> Accessed on: 11 August 2015.

CALDEIRA, Junia Marques. The Brazilian Square: trajectory of urban space - origin and modernity. 2007. 432f. Thesis (PhD in history) - State University of Campinas, Institute of Philosophy and Human Sciences, Campinas-SP. Available at in: <http://www.bibliotecadigital.unicamp.br/document/?code=vtls000428029> Accessed on: 26 August 2015.

CARLOS, Ana Fani Alessandri. The place in/of the world. São Paulo: FFLCH, 2007.

CORRÊA, Leon Alves; XAVIER, Gilson. Quirinópolis' urban development. In: URZEDO, Maria da Felicidade Alves (Org.) Quirinópolis, Mãos e olhares diferentes II, História e Imagem, 1932-2012. Goiânia: Kelps, 2012, p.185-194.

D'ABADIA, Assilvo. A summary of the history and origins of Quirinópolis. In: URZEDO, Maria da Felicidade Alves (Org.). Quirinópolis, Mãos e olhares diferentes, 19322010. Goiânia: Kelps, 2010, p.45-88.

DAMIANI, Amélia Luisa. Place and the production of everyday life. In: CARLOS, Ana Fani Alessandri (org). New paths in geography. São Paulo: Contexto, 2001. p 161172

FERREIRA, Luiz Felipe. Recent perceptions of the concept of place and its importance for the contemporary world. Território Magazine. Rio de Janeiro, v.5, n°9, p. 65^83, jul/dez, 2000. Available in: <http://www.revistaterritorio.com.br/pdf/09_5_ferreira.pdf> Accessed 17 August 2015.

GIL, Ana Helena Correa de Freitas; GIL FILHO, Sylvio Fausto. Geography of everyday life: a reading of Erving Goffman's socio-interactionist methodology. Ateliê Geográfico - Electronic journal. Goiânia-GO v. 2, n° 2, p.102-118 aug/2008. Available at: <http://www.revistas.ufg.br/index.php/atelie/article/view/4708/3969> Accessed on: 12 August 2015.

HESPANHOL, Rosangela Aparecida de Medeiros; MOREIRA, Erika Vanessa. Place as a social construction. Revista Formação. Presidente Prudente-SP, v.2, n^{0}14, p. 48-60, 2007. Available at: <http://www2.fct.unesp.br/pos/geo/revista/artigos/6_moreira_e_hespanhol.pdf> Accessed on: 20 August 2015.

HOLZER, W. Place in Humanist Geography. Território Magazine. Rio de Janeiro. v.4, n°7, p. 67-78. Jul/Dec 1999. Available at: <http://www.ppgg.ufam.edu.br/attachments/article/88/Espa%C3%A7o,%20Territ%C3%B3rio%20e%20Cultura%20(TEXTO%205).pdf>. Accessed on: 27 July 2015.

BRAZILIAN INSTITUTE OF GEOGRAPHY AND STATISTICS (IBGE). IBGE: Cidades@: Quirinópolis. Rio de Janeiro, 2014. Available at:

<http://www.cidades.ibge.gov.br/xtras/perfil.php?lang=&codmun=521850&search=||infogr%E1ficos:-informa%E7%F5es-completas> Accessed on: 20 August 2015.

LISBOA, Severina Sarah. The importance of geography concepts for learning school geography content. Ponto de Vista magazine. Viçosa-MG, v.4, n°4, p. 23-35. 2007. Available at: <http://www.coluni.ufv.br/revista/docs/volume04/importanciaConceitosGeografia.pdf> Accessed on: 05 August 2015.

MARTINS, Pollyany Pereira. The socio-spatial dynamics of Aragarças-Goiás: everyday life in the construction and structuring of urban space. 2015. 192f. Dissertation (Master's in Geography) - Federal University of Goiás, Jataí.

MASSAE, Denise. Slides 109, colour, 2012, based on ROBBA, Fabio; MACEDO, Silvio Soares. Brazilian squares. Landscaping: parks and squares. São Paulo: Editora da Universidade de São Paulo: Imprensa Oficial do Estado, 2002 -[Quapá Collection].

MATTOS, Georgides de Souza; PARREIRA, Airosa Martins. History of Quirinópolis. Fragments written in March 1988. In: URZEDO, Maria da Felicidade Alves (Org.). Quirinópolis, Mãos e olhares diferentes, 1932-2010. Goiânia: Kelps, 2010, p. 33-43.

MORAIS, Eliana Marta Barbosa de. The idea of nature in everyday practice. In: CAVALCANTI, Lana de Souza (Org.). Geografia da cidade. Goiânia: Alternativa, 2001, p.79-132.

NEVES, Paracy Corrêa. The formation of Quirinópolis' urban space: a historical possibility from 1832 to 2010. 2012. 98f. Master's dissertation. Pontifical Catholic University of Goiás, Goiânia, 2011.

PAGLIOTO, Iuri Freitas. Public Square: The Public Place as Cultural Heritage. An analysis of Gomes Freire Square in Mariana-MG. 2015. 61f. Monograph (Degree in Geography) - Federal University of Viçosa, Viçosa- MG. Available at: <http://www.geo.ufv.br/wp-content/uploads/2015/07/Iuri-Freitas- Paglioto.pdf> Accessed on: 26 August 2015.

SANGIM JUNIOR, Odir; SAGIM, Mírian Botelho. Quirinópolis History. Goiânia: O Popular, 2000.

SANTOS, Gilberto Celestino dos. Quirinópolis, development that consolidates the importance of the municipality. In: URZEDO, Maria da Felicidade Alves (Org.) Quirinópolis, Mãos e olhares diferentes II, História e Imagem, 1932-2012. Goiânia: Kelps, 2012, p. 29-34.

SANTOS, Milton. Metamorfoses do espaço habitado, fundamentos teóricos e metodológicos da geografia. São Paulo: Hucitec, 1988.

SEGAWA, Hugo. To the love of the public: gardens in Brazil. São Paulo: Studio Nobel: FAPESP, 1996.

SILVA, Marta Correa da. Handicraft places in the city of Quirinópolis - GO: Diagnosis of locally-based know-how. 2014. 57f. Monograph (Coursework) - State University of Goiás, Quirinópolis.

CHAPTER 5

TEMPERO DE BOTECO AND THE GEOGRAPHY OF TRADITIONS: IN THE URBAN SPACE OF QUIRINÓPOLIS (GO)

Silea Xavier Camilo Ferreira[38]
SOUZA, Edevaldo Aparecido[39]

Introduction

This text is an excerpt from a monographic study on pubs in the urban space of Quirinópolis, a municipality located in the south-west of Goiás and in the microregion of the same name, also known as microregion 018, in the state of Goiás. The monograph was presented and defended in 2015 for the degree course in Geography at the State University of Goiás (UEG), Quirinópolis Campus.

This micro-region is made up of eight other municipalities: Gouvelândia, Paranaiguara, São Simão, Cachoeira Alta, Caçu, Itarumã, Itajá and Lagoa Santa. Its neighbouring municipalities are: Bom Jesus; Cachoeira Alta; Castelândia; Gouvelândia; Paranaiguara; Rio Verde and Santa Vitória in the state of Minas Gerais (Map 1).

In every town in the interior of Goiás, there are small businesses selling drinks, food and products to homes and farms, which here are known as botecos. In this context, Quirinópolis is one of the most promising municipalities in the region and the state, not only because of its privileged geographical location, but also because of its economic figures. It is among the fastest-growing cities in the Centre-West region, especially due to its agriculture (soya and, above all, sugar cane) and livestock farming (PREFEITURA MUNICIPAL DE QUIRINÓPOLIS, 2009).

The theme "Tempero de boteco e a Geografia das Tradições: no espaço urbano de Quirinópolis (GO)" was chosen after observing people gathered in front of the city's traditional small businesses to relax with friends, drink a cold beer and eat some of the local delicacies. The aim was to identify the contributions these businesses have made to the town, as well as the strategies they have used to stay in the market as the number of bars and snack bars has increased over the years.

---

[38] Graduated in Geography from the State University of Goiás, Quirinópolis Campus. .**sileaxavier.20@gmail.com**
[39] Professor of Geography at the State University of Goiás, Quirinópolis Campus. .ediueg@gmail.com

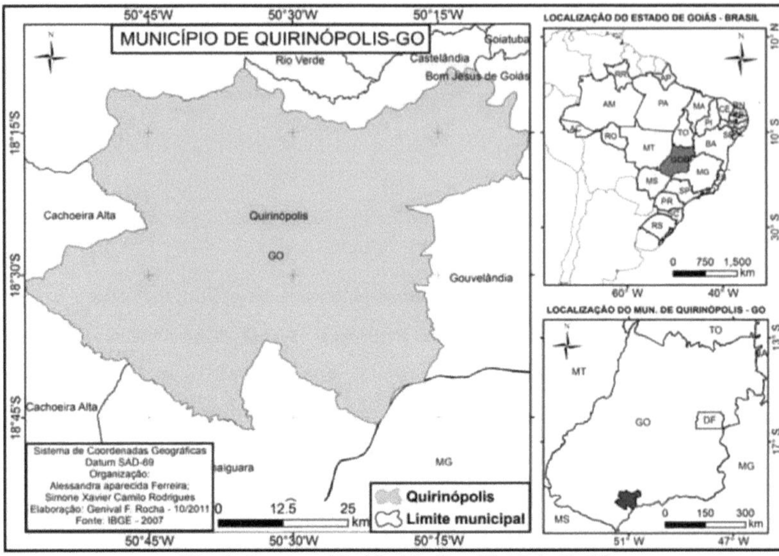

Map 1 - Location of the Municipality of Quirinópolis-GO
Source: IBGE (apud FERREIRA; RODRIGUES, 2011)

The botecos surveyed were the oldest in the city and those that still retain some of their traditional characteristics in terms of physical aspects, the organisation of the space and customer service, including the fact that they are known by name. We know that loyal customers are crucial for any business, but this doesn't happen overnight. It takes time, careful observation and effort to establish lasting relationships.

The aim of the research was to identify the contributions of these commercial establishments to the town, as well as the strategies used to stay in the market as the number of bars and snack bars increased over the years. The criterion for choosing the bars to be researched was *age* (as long as they had maintained some of their traditional characteristics in terms of physical aspects, the organisation of the space and customer service). Five establishments were chosen: Bar e Lanchonete do Allan, Bar Jovem Guarda (formerly Bar do Aparício), Bar Copo Sujo, Bar da Rodoviária, in Rua Rio Preto and Pastelaria Tradição - all in the urban area of Quirinópolis, as shown in Map 2.

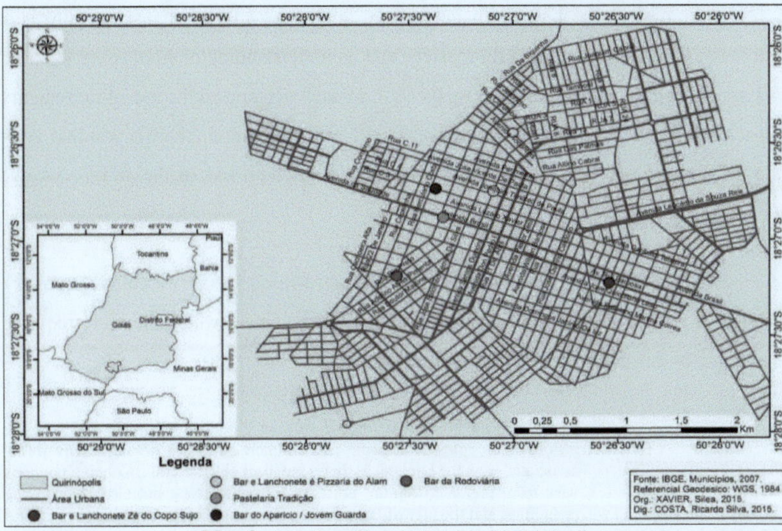

Map 2 - Location of bars in the city of Quirinópolis
Source: IBGE, Municípios (2010); Org.: FERREIRA, Silea Xavier Camilo (2015); Digitisation: COSTA, Ricardo Silva (2015).

As for the method used in the research, we adopted the phenomenological method devised by Husserl et al. (apud SADALA, 2014, p. 2), according to which it is necessary to get to know the subject by "returning to the lived world, to the world of experience, the starting point of all sciences". According to the author, this method proposes not explaining the phenomenon a priori, but describing it or looking for causal relationships, in other words, turning to the things themselves from how they manifest themselves.

Tradition and the art of cooking in geographical science

Tradition in the botecos of Quirinópolis is made up of the transfer of knowledge and customs from generation to generation, seeking to maintain the oldest and most characteristic aspects of each boteco's identity. This tradition, which is highly valued by the owner and family members of the boteco, as well as by customers, is noticeable in all the establishments analysed, from the oldest to the most recent, and the quest to maintain it is notable: be it in the menu, the physical structure of the building or the service.

It should be emphasised that what is meant by tradition is no longer its originality, i.e. some material and immaterial elements have already been modified, some modern technologies have been incorporated, but we can easily call it tradition re-signified, because many characteristics from the past have been retained in these establishments.

Activities in the kitchen in themselves qualify as an art because, within their processes, there is a need for skill, mastery and an understanding of what is being done. The art of cooking is differentiated by the skill of each person and the use of ingredients in each recipe, the secrets, the dedication, the determination and the affection that goes into the practice of know-how.An approach from the perspective of the art of cooking: some discussions about cookery

A simple concept, but one that is widely used in our country's culture, cookery consists of the act of cooking, considered by many to be an art, which therefore requires technique. In this context, it can be mentioned that

> Cooking is the art of cooking, i.e. making food, and it has evolved over the history of peoples to become part of their culture. Cooking is invariably associated with the kitchen, as this is the ideal place to cook. Cooking methods vary from region to region, not only the ingredients, but also the cooking techniques and the utensils themselves. For example, the cataplana is a typical cooking vessel in the Algarve, equivalent to the tajine in Morocco. Alheira de Mirandela is one of the most exclusive foods in Portuguese cuisine, while in Brazil, typical dishes include Brazilian feijoada and churrasco. The main cook is usually known as a *chef,* thus recognised for his good cooking and culinary skills (CORREIA, 2014).

According to Gastroconsult (2011), cookery often establishes links with other aspects of culture. With regard to religion, for example, there is a taboo among Hindus on the consumption of beef, while pork is extremely forbidden among Muslims and Jews. There are also certain political and social positions such as vegetarianism, which excludes the consumption of food from or derived from animals such as milk, cheese and eggs.

To this end, the art of cooking must be seen as something that is passed down through the generations, or just as an obligation, since eating is a physiological necessity for living beings. Some people use cookery to prepare their food at home in a simple way. However, others use it for commercial purposes, working on the economic side of this art and turning it into a profession.

Within his conceptualisation, Aurélio Buarque de Holanda Ferreira (2011, p. 137) emphasises that "cuisine consists of the art of cooking". There is a doctrinal variety in the sense of contextualising cooking as an art. From this perspective, Vainsencher (2007, p.1) claims that:

> Each people has its own type of cuisine, its own way of preparing food. From the point of view of folk culture, we can see that through different forms, mixtures, temperatures, smells and colours, people turn food into an attraction. To eat is to know, says an old saying. In this way, all the world's cuisines represent forms of knowledge. They are cultural signals transmitted through

> taste, sight and smell. They are tastes, sensations, textures or touches that whet desires.

In this way, we can see that cuisine is closely linked to the tradition and culture of a people. In each region, through tradition, specific eating habits can be seen, such as in the state of Goiás, where the consumption of pequi, a fruit native to the Cerrado, is common, as the local biome favours its reproduction.

This is not to say that everyone who lives in the area uses the fruit, but they do try it or have tried it at one time or another and, if they like it, they continue to consume it habitually, because in many households it is found on the tables as food and in a variety of dishes.

Another example of culinary tradition is in the southern region of Brazil, which consumes the largest amount of beef in the country, as it is a food that has traditionally been part of that region's cuisine. This is why it is said that cookery is closely related to the tradition of a people.

It is also necessary to emphasise the origins of Brazilian cuisine, as Vainsencher (2007, p. 1) explains:

> Brazil has an original and expressive cuisine. Over the course of 500 years, Brazilians have assimilated and transformed European cuisine, especially Portuguese cuisine, the spices that the colonisers brought from the East (China and India), adding ingredients from African and indigenous cuisines (that of the Indians of the Amazon and the Pantanal of Mato Grosso). The latter, a permanent feast of stewed fish, game and seasonal fruit, was already present when the country was discovered. All this without ever harming or jeopardising the stability of the environment. It was food exchanges, therefore, the union of different paths and life experiences, ethnicities and cultures, the miscegenation of tastes, forms and aromas, that generated a new and rich cuisine: Brazilian.

According to Vainsencher (2007 p.1) "manioc flour, food boiled or roasted in banana leaves, food made with corn, paçoca (fish or meat pounded and mixed with flour)" were inherited from the Indians. Also the moderate use of salt and spices, as well as "cooking with an oven and cooker, the use of ceramic utensils, the virtues of eating fresh food, and the foods seasoned by the hands of the native Indians. Without this, the national cuisine would be very poor today".

> As well as refinement, the Portuguese colonisers introduced some important ingredients to Brazilian cuisine: coconut (brought from India), salt and cinnamon powder mixed with sugar. Sarapatel, sarrabulho, panelada, buchada and cozido are not part of African cuisine, but Portuguese. The first two also came from India via the colonisers. The Portuguese confectionery brought: pudim de iaiá, arrufos de sinhá, bolo de noiva, pudim de veludo. There were

> also many Moorish and African delicacies, such as alfenim and couscous, and fruits such as mango, jackfruit, breadfruit and carambola, which were brought from the Orient. From the already famous Portuguese stew came the idea of including black or mulatto beans, meat and lots of vegetables to make a unique dish: feijoada (VAINSENCHER, 2007, p. 1).

The text above shows that our cuisine has received a strong indigenous influence, due to the naturalness of the food. Fish caught at the moment of consumption and steamed, the use of manioc that can become flour, giving flavour and dough to various types of food, such as tapioca, widely consumed in Bahia. The presence in our cuisine of spices brought from India by the Portuguese contributed to Brazilian seasoning, adding flavour to food, as well as salt, which was unknown to the Indians.

You can see that some of the foods used in our cuisine come from the colonisers and the indigenous people who inhabited the country before colonisation.

As mentioned above, India brought its contribution to cooking through spices, which were brought by the colonisers.

Black slaves, the product of trafficking by the colonisers, also contributed to the formation of eating habits. The African way of life and their traditions added flavour and spice to Brazilian cuisine. In relation to this African contribution, Vainsencher (2007, p. 1-2) explains that:

> The African presence on the Brazilian table has palm oil and chilli peppers (not the native ones, used by the Indians, but the malagueta, brought by blacks from Africa) as its greatest representatives. The palm tree from which the oil is extracted came to Brazil from Africa in the early decades of the 16th century. All the dishes brought from Africa were then reworked and recreated by Brazilians, who began to use palm oil and local elements. Although African, the yam was known in Portugal. Caruru, on the other hand, as it is known, is an African dish that has kept its indigenous name, but has acquired another content: chicken, fish, beef or shellfish. When they arrived in the country, black slaves were already cooks. Learning from the Portuguese and surpassing them with the diversity of spices she was able to handle, the Africans competed with the Indians for the secret of fine dining.

Such a wealth of African spices has contributed to the formation of Brazilian cuisine . Even though it was influenced by the Portuguese, Africans, Indians and Indians, the mixtures of spices made here in our lands resulted in a final product that can be categorised by region. Each region has acquired its own particular culinary tradition. With regard to this regional context, it can be inferred that:

> Today, each Brazilian region has its own typical dishes. In the North, due to the presence of forests, indigenous influence and generous hydrographic

> basins (the Amazon River and its tributaries, in particular), the consumption of freshwater fish predominates (acari, auanã, cascudo, surubim, pirapitinga, piranambu, tucunaré, tambaqui, pirarucu, tainha, camurupim, itui, jandiá, xaréu, curimatá, cangati, piranha, among others); cassava and fruit: açaí, tacaca, buriti, taperebá, ginja, pupunha, murici, uamari, cupuaçu, bacuri, camapu, uxi, angá, piquiá, camutim, cutitiribá, grumixama, cubiu, guaraná (VAINSENCHER, 2007, p. 2).

Cassava, fish and açaí have become important staples in the cuisine of Goiás and Quirinopolina. Cassava is widely consumed in the bars of Quirinopolis, mostly as an accompaniment to various dishes, such as fried fish and picanha on the grill, and it is often used to make starch, biscuits and cakes. It is an important ingredient in Goiás cuisine, both as a side dish and as an ingredient for other menus.

Fish is also very popular with customers in commercial establishments, whether fried or roasted on the tile, both forms of preparation are always a popular dish in pubs, as is fish broth, which is especially appreciated on cold days.

In addition to the traditional foods that are common to the population of Quirinópolis, other types of food that enrich Goiás cuisine are being added over time, such as açaí, the consumption of which is growing every day in the city of Quirinópolis, both in ice-cream parlours and in specific establishments selling the product.

Moving on to northern cuisine, Vainsencher (2007, p. 1) states that, as well as being tropical, it is also ecological and comes with a wide variety of chillies: "cajurana, mata-frade, murupi, camapu, murici, olho-de-peixe, ova de aruana, pimenta de-cheiro and olho-de-pomba".

> Many other delicacies are eaten in this region: maniçoba, jaraqui stew, duck in tucupi, grilled tambaqui, tacacá gourd, mujanguê (the famous turtle egg farofa), various types of turtles (juruá-açu, capitari, tracajá, matamatá, cabeçudo, pitiú), bacuri and cupuaçu creams. In the north-east, you'll find dishes based on beans, yams, cassava (called aipim in the south of the country), coconut milk, palm oil, fish, crustaceans and native fruits. There are many delicacies in the region: buchada, sarapatel, arroz-doce, tapioca, caldo de cana, as well as regional fruit sweets and/or ice creams: papaya (green), guava, cashew, pine cone, sapodilla, banana, tangerine, mangaba, coconut, mango, umbu, jackfruit, pineapple, araçá.In the south and south-east, where there are large herds of cattle and sheep, people eat barbecued meat and sausages, accompanied by white rice, mayonnaise salad, toasted manioc flour, boiled cassava, green salads and bread. Other traditional dishes include stew on a stick, ox atascado, lamb shank, ribs, sheep barbecue, tripe grossa, and others. The gauchos, in particular, consume a lot of chimarrão, a hot tea made from crushed bitter mate leaves (VAINSENCHER, 2007, p. 2).

It can therefore be seen that climate and vegetation, among other environmental and historical factors, influence the eating habits of certain places and/or regions. To this end, we realise that not only the culture brought by colonisers has interfered in the final product that is Brazilian cuisine. All the components together contribute to the flavours of our country.

History of Quirinópolis' pubs

At first glance, pubs appear to be a retail outlet like any other, i.e. where drinks are bought for consumption on the premises and/or to take home. However, over the years, these places have become a meeting place for people after work and this increase in demand has led to the need for owners to offer an additional attraction. This is how aperitifs and drinks came to be. By socialising, people have references, friendships and bonds that become part of their lives through memories. According to Kenski (1998), the bar is a spatial mark that recalls past events and fixes an era in the memory: "inside, the bar looks like so many other botequins: a modern Formica counter, chairs and tables".

There has always been a masculinised view of bars and pubs due to the presence of alcoholic drinks and, in most of them, snooker. Over the years, women have started to frequent these places and today the presence of women in these venues is commonplace.

It can be seen *on the* spot that it is not just a place for shopping, but also a place for meetings, fun, leisure, relaxation and informal conversations, regardless of where it is located, whether on the roads, on the outskirts or in the city centre. We realise that, even with the passing of the years, botecos traditionally still preserve their main characteristics as a cultural essence, in their forms and contexts.

There are establishments that try to maintain the typical local dishes and speciality dishes that are specific to the restaurant. This way of maintaining traditional dishes becomes attractive, so that even over the years the restaurant maintains and retains its clientele.

The term "boteco" originated in Portuguese and means "botica", from the Spanish "bodega", which in turn comes from the Greek *"apotnéke"*, which means "warehouse", "drinking house" or "shop" in which goods were sold at retail. According to Aurélio (2009, p. 264) "bar is the counter in front of which people stand or sit on high stools to consume drinks and light delicacies." The same dictionary states that "botequim is the commercial establishment where drinks in general (alcoholic beverages, soft drinks, coffee, etc.) and small snacks are served". It also states that a boteco is a "tent that is set up around the stalls at fairs".

For Silva (2009), the word "boteco" still has a derogatory meaning (which produces effects of meaning that feed inequality), as something without value. Therefore, the identity of the person who goes to the boteco is different from those who go to the bar and the botequim.

In Brazil, botecos (or the popular "buteco" or "butiquim") have become known as a place that serves as a meeting place for "bohemians" looking for a quality drink, such as the famous "cachaças de engenho", an artisanal drink served in portions, as well as the consumption of cheap snacks and relaxed conversation with beer (SILVA, 2009).

In many Brazilian towns, the most attractive snacks are the "simple" ones. Among the most peculiar delicacies in the botecos, we can mention the liver served as a "tira-gosto", the spicy chorizo or the fried jiló, accompanied by the famous "loira gelada", also known as draft beer or beer, and also the caipirinha, or the famous cachaça.

There is even a "competition" between the bars when it comes to snacks. Thus, some offer typical local dishes on their menu, in order to provide an extra attraction for customers. You can see traces of Quirinópolis' gastronomic culture in the local boteco flavours, especially in the typical dishes of the region, such as rice with pequi, chicken with guariroba, pastries made with shredded chicken and guariroba, and especially the empadão goiano. There is also the local dish known as "Chica Doida", created by Mrs Petrolina Ferreira Cabral (in memorian) in 1950, which arose from an improvisation when she and her husband made pamonha for their friends (ANDRADE; SILVEIRA, 2010).

Petrolina recounts that the dough yielded more than expected and the straws were few and far between, so her husband suggested making baked pamonha with "lots of chilli peppers, cheese, sausages, jiló and, as it was pequi season, to add it too" (ANDRADE; SOUZA, 2010, p. 648). The dish was so tasty that it became a hit, giving rise to Chica-Doida, in honour of Francisca, an employee of the house at the time, who was "crazy" about eating chilli peppers.

In 2008, Senhora Petrolina was honoured at the Quirinópolis Gastronomic Festival and Chica Doida (50 years later) was recognised as a typical dish of the town, as it is much appreciated by the population at school parties, dinners, June festivals and other events. With the success of the Festival and media coverage, especially on television programmes, Chica Doida became known throughout the country and people from various states visit the city to enjoy this dish with its unparalleled flavour (ANDRADE; SILVEIRA, 2010).

It's a fact that Brazilian cuisine is rich, diverse and tasty, because there is an infinite variety of spices that, used individually or mixed with others, give dishes a special flavour.

The following subtopic presents the history of Quirinópolis' oldest bars/booths, as well as the origin of their owners, what motivated them to come to the town; it also deals with the products they sold at the time and, finally, what this market segment represents for their family and for the population of Quirinópolis in the past and today.

Geographical research looks at the pubs of Quirinópolis: describing traditional businesses

The bars located in the urban area of Quirinópolis stand out for their importance, as they are part of the daily life and history of the place. It is understood that urban space "tends to go beyond the limits of physical agglomeration (the city) and concentration through the tactics and strategies of different capitals and labour to maximise their respective mobilities" (LIMONAD, 2007, p. 147).

Costa (2010) says that when approaching the category of **space**, it should be considered as a set of social relations realised through forms that bear witness to a history written by processes of the past and present, because space is defined as a set of forms that represent past social relations and current ones. From this point of view, space is therefore:

> [...] one and multiple, through its various parcels, and through its use, it is a set of commodities, whose individual value is a function of the value that society, at a given moment, attributes to each piece of matter, that is, each fraction of the landscape (SANTOS, 1997, p. 85).

Pubs are part of the historical and contemporary times of Quirinópolis' urban space, in other words, they existed in the past and still exist, maintaining historical cultural traits interwoven with others from modern times. In this context, Limonad (2007, p. 161) explains that "space and history cannot be dissociated and both are inextricably linked to social life, material conditions and the development of productive forces and the technical-scientific environment".

Still according to the author (LIMONAD, 2007, p. 148), pubs can be understood as "urban practices and relationships (the urban way of life)". This discussion sheds light on the issue of traditional businesses:

> [...] they integrate the Quirinopolina micro-region culturally, incorporating other local elements such as cuisine and religious festivities into this context, presenting the communities as part of the same cultural compartment and expressing their regional identities through the ways of life and values that form the families of the interior of Goiás; these mixtures broaden the attractions of the event (SANTOS, 2010, p. 208).

Pubs are part of urban memory, as many have disappeared and today only "[...] evoke memories, emotions, impressions" (MORAES, 2012, p. 83). The existence of these small businesses in places promotes "the maintenance of the community's living history" (CORIOLANO, 2009, p. 153) and can contribute to the preservation of a form of commerce that no longer exists in modern businesses, becoming an art that generates or complements income for families who still make a living from this activity.

Using Bezerra (2013, p. 118), it can be said that the boteco is "a symbol of social and cultural representation of the community" in Quirinopolina. In order to highlight these social representations, we need to:

> [...] to feel and let oneself be invaded by a sense of discovery. It is a singular imaginary universe, a reflection of a unique social scenario, culturally constructed over many decades as a representation of a people, of a different but original society. (BEZERRA, 2013, p. 125).

It can be said that researching the pubs of Quirinópolis is to bring out the feeling of discovery of a people whose traditions are rooted in the place, with their own way of life and cuisine.

In view of the above, the research revisits the late-night conversations that took place in these places, accompanied by a cold beer and typical local dishes - portraying their importance as a place of leisure. In addition, we set out to demonstrate the strategies used by these bars/botecos to remain in the market to this day, since we believe that the fact that the owners of these establishments know the names and habits of their customers (patrons) is a great incentive for them to become regulars. So the first step to winning them over is to keep their name. It can be seen that, in the establishments surveyed, the owner greets his customers by name, knows their preferences and treats them like friends.

According to Cabral and Souza (2010, p. 352) the cuisine of Quirinópolis is very diverse and "the flavour and aromas characteristic of the dishes of our ancestors are still very much appreciated today by all social classes and age groups for the seduction and pleasure they offer".

pequi, guariroba, chica doida, pamonha and frango caipira ao molho are typical dishes of Quirinopolina's cuisine, and are also enjoyed when families and friends get together at home and/or in the city's pubs.

With this in mind, this study looked at part of the history of the pubs Bar e Lanchonete do Allan, Bar Jovem Guarda (formerly Bar do Aparício), Bar Copo Sujo, Bar da Rodoviária and Pastelaria Tradição, with the aim of rescuing the history of these establishments, as well as demonstrating their importance to the urban and cultural development of the municipality.

Bar Universo was founded in 1973, known as Bar do Aparício, a place frequented by people who wanted to eat fried pork skewers (the most popular snack at the time) and have a cold beer, cachaça or other drinks sold in portions. According to the interviewee, the bar was rented out to a former employee after the car accident that led to the owner's death (REZENDE; PASQUALI, 2012).

Rezende and Pasquali (2012, p. 294) state that Aparício's Bar was well known in the city. There were different types of drinks, mainly shots, "pool tables, two horizontal counters, one refrigerated for beer and soft drinks, the other for dry goods". In their research, the authors describe what their interviewee said about the establishment.

> A rented building that opened its doors between 6am and 7am with customers waiting to have coffee and eat fried biscuits, cheese bread and or pastries. It was busy all year round and stayed open until 4am and we slept very little. There was a "reserved" area for families to feel more at ease and another area for pool tables, with two **billiards** tables **and people playing 'Jogo Vida'. It was a bar** frequented by many people from the city (REZENDE; PASQUALI, 2012, p. 294).

Mrs Olina, Mr Aparicio's widow, said that the place was frequented by farmers, shopkeepers and rural workers. It was family-run and had the help of an employee. The products sold were: snacks, sausages, olives, peas, salami, pork, fresh cheese, pastries, quibbles, cheese bread, cakes, pizzas and fried wheat flour biscuits. The snacks were kept in the shop window or greenhouse and in the evenings, pork skewers with manioc flour were served. Figure 1 shows Mr Aparicio and his family inside the bar, in front of the walls and cupboards used to store the drinks.

With regard to the establishment, the interviewee explains that the place represented a lot to the Quirinopolina community; it was a place for leisure, chatting and doing business. She emphasises that the building still preserves part of its physical structure, such as the table, the floor, the ceiling and the doors, which are "in the colonial style, made of wood from joinery"[40] .

[40] Interview account given by Mrs Olina Maria da Silva e Souza.

FIGURE 1: Mr Aparicio's family and the interior of his Boteco
Source: Souza (apud REZENDE; PASQUALI, 2010, p. 294)

The physical structure of the establishment has not changed over the years. Some changes are only noticeable in the sign. The traditional features of the old Bar do Aparício, now the Bar Jovem Guarda, have been preserved (Figure 2). Inside the bar, two kettles are used, one old and the other modern.

FIGURE 2: Layout and current façade of the Bar Jovem Guarda
Source: FERREIRA, Silea X. C. (2015)

Another business in the same segment that also stood out in the 1970s and is still active today is Allan's Bar, which according to Rezende and Pasquali (2012, p. 295), was founded in 1971 under the name Bar Goiano. It belonged to "five former farmer brothers who, after having their land 'taken' by the waters of the São Simão-GO hydroelectric dam, moved to the city of Quirinópolis and bought the bar, located on Avenida Mato Grosso (Garibaldi Teixeira)".

In 1977, the business moved to Rua José Joaquim Cabral, where it remained until 1985. When it was renamed Lanchonete Flamboyant from 1986 to 1988, the place and name were changed again. What set this bar apart from the others was the sale of bauru with a snack[41] , which won over customers' taste buds. The owner at the time said

> [...] bauru com quitute [sic] was Bar Goiano's launch, we didn't know if it was available in any other bar and it soon won over the customers' taste buds. Another good-selling snack was the round cake made in small moulds for individual portions, baked in several moulds at once and for the baby chocolate - milk with toddy or guarana. (REZENDE; PASQUALI, 2012, p. 295).

It can be seen in the story that Bar Goiano also preserved local traditions and that the word "freguês" was used to refer to customers. Almost all of the bars had the same characteristics, snacks, drinks and delicacies, with only a few differences that made the place a favourite for some of the people who frequented it. In the case of the Goiano Bar, the speciality was the bauru with snacks.

The authors Rezende and Pasquali (2012, p. 295-296) also comment that the owners of Bar Goiano sold it with a view to changing their business. In 1977, they opened Lanchonete Pindorama, "a bar with a new look, similar to those in larger urban centres". In 1986, the business opened a branch on Av. Adolfo José D'Abadia, as shown in figure 3.

Figure 3: Pindorama Lanches, a leisure centre built in 1986
Source: REZENDE, N. A. (2012)

41 Quitute is a cooked, tinned ham-like product used as a filling in bauru by the owners of Bar Goiano in the 70s.

They add that:

> The Pindorama snack bar only closed its doors on Thursdays and Fridays during Holy Week. On Hallelujah Saturday, at around 6.30am, the doors were reopened to welcome customers. At weekends, especially on Sunday afternoons, families would gather there for ice cream (Kibon). On that day, if the Municipal Championship was over, the players from the Clube de Regatas São Francisco would also gather there to celebrate being champions of the Municipal Tournament or other tournaments (REZENDE; PASQUALI, 2012, p. 297).

The aforementioned authors (2012) also report that Lanchonete Pindorama was one of the places where friends met, warmed up for dances, carnival parties at the Itamaraty club and celebrated victories in sports championships. The brothers were responsible for serving the public and the sisters for the cooking, but one of the brothers specialised in preparing the batatinha, torresminho, picadinho de fígado and filé a palito.

The authors (2012) emphasise that working in a cafeteria requires planning, dedication and the involvement of everyone who works there. It is tiring and rewarding at the same time, as it is a social and emotional place where people meet to have ice cream, drink, eat, relax and chat. It's also where the families who work there earn their living. It requires commitment in terms of:

> [...] adequate labour; attention to drinks and food, as well as practically daily shopping; work, often beyond opening hours; determination and flexibility to deal with difficult or inconvenient situations; eyes focused on results that imply customer satisfaction (REZENDE; PASQUALI, 2012, p. 298-299).

Still on the subject of the trajectory of Allan's Bar, in July 1986 the Pindorama Snack Bar, located on Avenida José Joaquim Cabral, was sold and in August of the same year the brothers opened the Flamboyant Snack Bar, "a space organised with 64 sets of tables, 4 horizontal freezers for beer, a refrigerated counter for soft drinks and a total of 12 employees" (REZENDE; PASQUALI, 2012, p. 299). However, according to the authors (2012, p.30), the instability of the country's currency, the increase in interest rates and the replacement of the Cruzeiro by the Cruzado jeopardised the family's financial health, making it necessary to give up this activity and remain only with the snack bar that already existed, keeping "the traditional name Pindorama Lanches, also known as Pizzaria e Lanche do Allan" (Figure 4).

According to Mr Allan, the building was constructed in 1984 on a plot of land owned by the family, close to Quirinópolis Hospital and commercial outlets. It was considered an ideal location, with good opportunities for the snack trade. According to the interviewee, the establishment has kept its physical structure preserved and has a permanent kitchen assistant who works 8 hours a day.

FIGURE 4: Current photo of Allan's Bar and Pizzeria, although the fancy name Pindorama Lanches is still on the façade

Source: FERREIRA, Silea X. C. (2015)

The products currently sold are: "drinks (beers, soft drinks and juices), savoury snacks (pastéis, pizza, coxinha, carninha, quibe, bauru) and others to the customer's taste"[42] . Some products are displayed in the greenhouse and can be served on plates with cutlery or simply handed to customers, who receive them with napkins.

According to the interviewee, the bar's customers are: labourers, civil servants, self-employed professionals (doctors, lawyers, shopkeepers, tradesmen), rural workers and farmers[43] . He also emphasised that he doesn't know how many loyal customers there are, but his clientele spans several generations, children and grandchildren of customers, in other words, since Bar Goiano was founded over 44 years ago.

One of the owners emphasised that the establishment represents a place where the community can get quick snacks during the week and, at weekends, it is a place to relax, ideal for having a beer and chatting with friends.

Another establishment surveyed was Pastelaria Tradição (Figure 5), which is both a bar and a snack bar, but with the difference that it doesn't sell alcoholic drinks, hot drinks served in portions, or even cold drinks, the popular beer.

[42] Interview report (2015).
[43] Interview report (2015)

Figure 5: Panoramic view of Pastelaria Tradição
Source: FERREIRA, Silea X. C. (2015)

The establishment has been operating since 1980. Its first location was on Avenida Brasil, opposite the current Calçados Castro shop, where it operated for 22 years. It currently belongs to his daughter and son-in-law, who together have continued the work of Mr José Ramos de Oliveira .[44]

The place still preserves traces of the original structure, such as the hipped roof and the concrete benches. The pleasant atmosphere is frequented by people from all walks of life.

According to Mr José Ramos, his pastéis are part of Quirinópolis' culinary tradition. He adds that "there is no better pastry and that many Quirinopolians have been eating pastries since their mothers' wombs". He also said that the first pastry shop in the city was his, which is why he was awarded the diploma of citizen of Quirinópolis by Mayor Sodino Vieira de Carvalho on 7 December 1988.

According to the interviewee, the filling options for pastéis are: meat, meat with cheese, ham with cheese and chicken with guariroba. Among the other products sold are coxinha, risole de carne, quibe, bauru, empanado de frango, empadinha, esfirra, queima língua, enroladinho, pizza, soft drinks and natural orange juice.

Worth mentioning is the Bar da Rodoviária (Figure 6) which, in 1993, began its activities selling industrialised dry and wet products and fried snacks, which are only resold as they are not manufactured on site.

[44] Report of an interview with the daughter of Mr José Ramos de Oliveira.

Figure 46: Bus Station Bar, a hangout for loyal customers and passers-by
Source: FERREIRA, Silea X. C. (2015)

In the landscape, you can see the traditional way: the owner uses the pavements to place crates of drinks, tables and a hand-written information sign with the brands of beer that the establishment sells. This establishment has always existed in the city's old bus station; there is a public payphone, men's and women's toilets, which, like all establishments, have an old structure whose paintwork is renewed once a year.

The working relationship is a family one, with the help of his children. There is a friendly relationship with the customers, who buy frequently and are local residents. The owner believes that the establishment represents a great deal for Quirinopolina's tradition, since there is a bus stop for boarding and disembarking, so some people can have water, a snack and buy other products before travelling, especially when they have children.

Also noteworthy among these bars is Bar do Zé, better known as the "Dirty Glass Bar" (Figure 7), due to the fact that this establishment serves its beer with a dirty glass of lemon and salt (Figure 8).

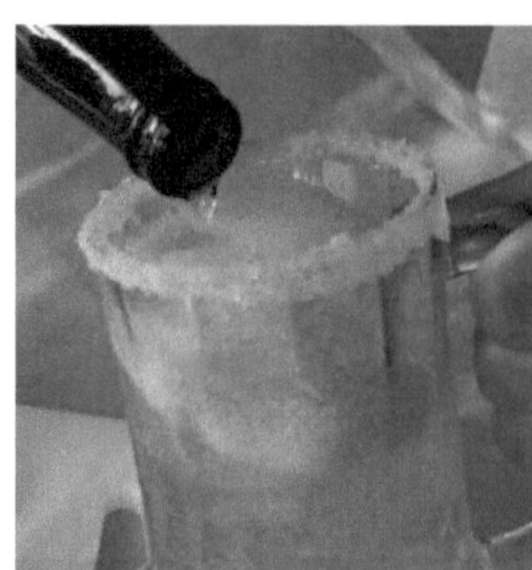

Figure 7: Current façade of Bar do Zé - Copo Sujo
Source: FERREIRA, Silea X. C. (2015)

FIGURE 8: Salt and lemon the difference at Copo Sujo Bar
Source: http://migre.me/vA3Dv (2015)

At the Copo Sujo pub, the kitchen is very simple, with a four-burner cooker, a fridge and a worktop for handling the spices that will make up the dishes. Bar do Zé's best-known dish is "fried fish", which is served with tomatoes, manioc and a cold beer. It should be emphasised that the fish is seasoned only with salt, passed through the egg and breaded in cornmeal, then fried in hot oil (Figure 9). Other dishes are served, such as: jantinha, broths (corn, chicken, beans, fish and mixed), picanha na chapa, chica doida, fried savouries, feijão tropeiro and others, accompanied by drinks (beer, soft drinks, juices and others).

FIGURE 9: The famous fried fish with its accompaniments at Bar do Zé
Source: FERREIRA, Silea X. C. (2015)

Through this research, it was possible to observe that the culture of frequenting the bars surveyed is passed down from generation to generation, with the traditional practices of each place remaining, making the social relationships established there important elements for reaffirming their ways of life.

Final considerations

In the course of this study, it became clear that the pubs surveyed are part of the daily life of some of the city's residents, who, when they leave work or at weekends, go there to eat, drink, chat and have fun with family and friends.

The results of this research show that the physical structure of the establishments, their cultural traditions, their history and their values have been preserved over the years, so that they have become a reference point in the city and ensure income for the families who dedicate themselves to this activity.

The survey also showed that the interviewees have a "little secret" in their seasonings, which makes their dishes unique, tasty and an invitation to those who haven't tried them yet. The dishes, savouries and snacks are simple and creative; the homemade seasoning and the way they are prepared are distinctive. The use of green smells and chilli brings out a special flavour.

It became clear that the establishments surveyed are part of Quirinópolis' history and are leisure centres frequented by people of different ages and social levels.

It was also noted that the customers in these bars are friends of the owner and that many of them have been going there since they were little, feel comfortable in the atmosphere and therefore bring their family and/or friends.

At the end of this study, it was possible to realise that the relationship of friendship, reliability and quality service have contributed positively to this type of business enduring for so long. Time and space are essential categories of analysis, as they are concepts that reveal transformations in local society. Thus, the permanence of these establishments takes place over decades and these traditions specialise not only physically (some of them have even changed location) but also culturally, given that flavours, especially from Goiás and Minas Gerais, merge to create dishes specific to each boteco.

Spatially, we still need to analyse the fact that these establishments, with the exception of Bar do Zé, are still located in the oldest part of the city, or close to the old centre of Quirinópolis. This spatial factor contributes to understanding the loyalty of many customers, as these places are home to the memory, imagination and history of many of these families.

In this sense, the objectives, questions and hypotheses of this study have been addressed in the final results of the research. By revisiting the history of the botecos, it was possible to prove their importance to the cultural and economic development of the urban area and the municipality as a whole. In the same way, the hypothesis and the answer to the initial question "What factors contribute to the permanence of these botecos in today's market?" are consistent with the spices and forms of organisation and service of these establishments, as the composition of the differential to please and retain their clientele.

Finally, this study has produced satisfactory results that can be used as a source of research for academics, traders and anyone else interested in learning about the spatial and cultural organisation of traditional bars in Quirinópolis. It is a contribution and a starting point for the development of further work, since the subject is not exhausted by this research.

References

BEZERRA, N. A. P. Representations of the environment in the imagination of Capim Dourado artisans in the municipality of Mateiros - TO. In: CORCINIO JÚNIOR, G. F.; SILVA, V. C. P. da. Nature and imaginary representations. Curitiba: Appris, 2013.

CORIOLANO, L. N. M. T. Arranjos produtivos locais do turismo comunitário: atores e cenários em mudanças. Fortaleza: EDUECE, 2009.

COSTA, E. B. A Concretude do Fenômeno Turismo e as Cidades Patrimônio Mercadoria: uma abordagem geográfica. Rio de Janeiro: Livre Expressão, 2010.

IBGE, Brazilian Institute of Geography and Statistics. Cidade@: Brazil municipality by municipality. 2010. Available at: <http://www.ibge.gov.br>. Accessed on: 20 September 2015.

LIMONAD, E. Urbanisation and the organisation of space in the age of flows. In: SANTOS, M.; BECKER, B. K. **Territórios, territórios:** ensaios sobre o ordenamento territorial. Rio de Janeiro: Lamparina, 2007. p. 145-170.

MORAES, G. M. C. Memory and history of portrait painters in Quirinópolis, Goiás - 1935 to 2012. In: URZEDO, M. da F. A. **Quirinópolis:** mãos e olhares diferentes II (1832 - 2012): história e imagem. Goiânia: Kelps, 2012. p. 83-110.

REZENDE, N. A. P.; PASQUALI, R. M. Recovering the history of bars and snack bars in Quirinópolis: 1965 to 1988. In: URZEDO, M. F. A. **Quirinópolis**: mãos e olhares diferentes II (1832 - 2012): história e imagem Goiânia: Kelps, 2012. p. 291-301.

SADALA, M. L. A. **Phenomenology as a method for investigating lived experience: a perspective on the thinking of Husserl and Merleau Ponty**. 2014. Available at: <http://www.sepq.org.br/iisipeq/anais/pdf/gt1/12.pdf>. Accessed on: 25 October 2015.

SANTOS, M. **A natureza do espaço**: técnica e tempo, razão e emoção. São Paulo: Hucitec, 1997.

SANTOS, J. C. V. **Regionalisation policies and the creation of tourist destinations between Lago de São Simão and Lagoa Santa in Baixo Paranaíba Goiano**. 2010. 366 f. Thesis (PhD in Geography) - Institute of Geography, Federal University of Uberlândia, Uberlândia, 2010.

CHAPTER 6

RUA RIO PRETO IN THE CENTRAL SPACE OF THE CITY OF QUIRINÓPOLIS: commercial tradition and merchant subjects

Renata Dias Borges Clemente[45]
SANTOS, Jean Carlos Vieira[46]

Introduction

The socio-economic transformations that took place in the last three decades of the 20th century led to changes in the shops that exist in the urban spaces of small towns. In this sense, this chapter aims to analyse the traditional shops on Rua Rio Preto in the city of Quirinópolis, Goiás, in order to understand the social and economic relations that still exist, bearing in mind the provision of new spaces for commercial reproduction in the Quirinópolis intra-urban area.

From this perspective, the research sought to analyse and answer the following questions: What are the traditional commercial relationships? Who are the people (customers/consumers) who remain loyal to the shops on Rua Rio Preto? How is this trade seen in today's urban dynamics? The study area is located in the oldest part of urban Quirinópolis, considered a commercial sub-centre of the city. According to Lopes (2011, p. 63), such urban areas "[...] are generally characterised by their power to attract consumers from all regions of the city".

This work is justified by the lack of research in the Geography course at the State University of Goiás (UEG) - Quirinópolis *Campus* and other universities looking at the traditional commerce of Rua Rio Preto, with its bars, snack bars, warehouses and the old bus station. In order to achieve the proposed objectives, the research starts with a discussion of the geographical category of *space*, arriving at the spatial organisation of the subjects involved with the old commercial enterprises in this part of the city. According to Limonad (2007, p. 161-162), in this urban spatial organisation, the:

> [...] relations of class and production, of domination and hegemony, do not exist on their own and their reproduction does not take place in a de-territorialised and a-spatial world. These relations materialise in space and structure it throughout history as territorial disputes of a political nature or economic and social inequalities.

Limonad (2007, p. 161) also points out that urban space and its "[...] history cannot be dissociated and both are inextricably linked to social life, material conditions and the

---

[45] Graduated in Geography at the State University of Goiás, Quirinópolis Campus. .rerenatinha7@hotmail.com
[46] Professor of Geography at the State University of Goiás: .svcjean@yahoo.com.br

development of productive forces and the technical-scientific environment". This landscape has another particularity that caught the eye during the field research: its built heritage is punctuated by 20th century architecture.

Theoretical references (books, newspapers, internet and magazines), fieldwork, photographic surveys, questionnaires and office work were fundamental to the construction of this work. According to Braga (2011, p. 134-135), fieldwork has been part of the strategies for producing geographical knowledge since its inception "[...] in the mid-19th century. As in school and/or academic practices, it has been appropriated by almost all conceptions of teaching or science, especially the empiricist currents".

In geography, Moura and Silva define field research as field trips which, in turn, are "[...] research that has contributed to understanding the socio-spatial relations produced by the segment and, with this, a better way of using the soil, the environment and the various human resources" (MOURA; SILVA, 2009, p. 16).

For Oliveira and Bueno (2009), reading landscapes with potential for fieldwork is important:

> [...] for students of Tourism, Geography, Primary Education, Secondary Education, Technical Education and various areas of knowledge, as it involves a multiplicity of educational actions and practices, which can take place both within the area itself and in interaction with others, as is the case of Geography with Biology, Literature, Architecture and others that work on different landscapes (OLIVEIRA; BUENO, 2009, p. 49).

Braga (2011, p. 136) states in his work that "[...] observation and data collection in the field must go beyond empiricism by systematising and articulating dialogues between the knowledge dealt with throughout the experience". Correlation, interpretation and analysis of the data collected were carried out in the office. It is hoped that the discussion about the historical and traditional commercial content of Rua Rio Preto, in Quirinópolis, can become a fertile field for new approaches and dialogues about the preservation and revitalisation of old urban spaces, as well as the enhancement of relations with different economic, social, cultural and environmental sectors.

Urban space: conceptual starting point

Observing the urban space allows us to see the spatialisation of the different social classes - deteriorated areas, segregated areas, noble areas or areas in the process of appreciation - as well as the historicity of society and its commercial areas, such as Rua Rio Preto, in the oldest landscape in the city of Quirinópolis. In this context, there is a dialogue with Costa (2010, p. 21), because when dealing with the category of space:

> [...] it should be considered as a set of social relations realised through forms that bear witness to a history written by past and present processes. Space is defined as a set of forms that represent past and present social relations.

Thus, it is important to discuss the concept of urban space in this work, because the production of capitalist urban space has a logic in the need for agglomeration that capital presents, "[...] but also in the need to hide social contradictions. This has resulted in different places, places for different classes and different groups, contradictory places" (CAVALCANTI, 2001, p. 16-17). Still according to Cavalcanti (2001, p. 14), this urban space:

> [...] it is the visible aspect of space, its formal, apparent expression. As a formal dimension, it expresses the content, the social relations that form it. It is therefore historical, social and concrete. Space is content, it is social relations in motion and which materialise spatially.

Observing urban space allows us to understand and perceive the current commercial spatialisation of Rua Rio Preto without, however, denying its tradition and history. According to Costa (2010, p. 23), understanding the production of space should not be restricted to the realms of politics, economics or culture, "[...] but to the interweaving of the three approaches for a better understanding of social spatialities, the economic and social reality of places at different historical moments". In view of this, it's worth remembering that urban space is:

> [...] the place where the reproduction of the family, the labour force (working class) and the social relations of production (goods of production) take place, which historically are concatenated to different degrees and intensities according to the stage of development of the productive forces, mainly concerning the development of the technical-scientific environment (LIMONAD, 2007, p. 148).

Thus, urban space is "[...] a real force field whose acceleration is uneven" (COSTA, 2010, p. 21). Lopes (2011, p. 61) characterises this type of space, in the context of its central area, as a "[...] densely occupied area, not allowing the installation of new commercial enterprises, in addition to the increase in land value, which also hinders the access of new commercial activities to this area".

Based on Damiani (1999, p. 161), we can see that it is necessary to incorporate the critique of everyday life, which emphasises the reproduction of social relations, into mobility routes such as Rua Rio Preto, in the urban space of Quirinópolis. Daily life becomes an important level for analysing reality when social reproduction fully affects the reproduction of life."

In this way, the ways of living commerce in the urban space and the customs of the people are understood in an absolute way, however rigorous the reproductive processes involving each local business may be. In this sense, the daily life of Rua Rio Preto is also created by the existing shops, but it is not alien and "[...] not marginal, although it is unequal to other sectors, moments and situations [...]. Daily life, as a simulacrum of full life thus modelled, allows the mode of production to function" (DAMIANI, 1999, p. 163).

Daily life, in relation to the economic and political spheres, expands the universe of analysis to other relationships between individuals and the urban space, "including particular, local ones. It includes what is lived, subjectivity, emotions, habits and behaviour" (DAMIANI, 1999, p. 163). In fact, the urban space of Quirinópolis, specifically Rua Rio Preto, is defined by everyday life and is part of the history of local society not only because of its economic and political aspects, but also because it was once a reference point for arrivals and departures in the city.

In the debate on urban space, which is expanding as geographical barriers implode and new readings emerge, authors such as Costa (2010) and Carlos (1999) emphasise that urban changes are inexorably invading people's lives. For the common man, this means the imposition of new behaviour patterns, new values, a new aesthetic. In this sense:

> [...] objects, bricks or houses are no longer sold, but cities. This means that space becomes a commodity, it enters the circuit of exchange, and with it previously unoccupied spaces become commodities, entering the sphere of commercialisation. (CARLOS, 1999, p. 175).

According to Carlos (1999, p. 175), "[...] geographical space articulates two dimensions, that of location (a point on the map) and that which gives content to this location, which qualifies and singularises it". This content is determined by the social relations that are established there - which gives space the characteristic of a social and historical product.

Because it has an indisputable materiality, the spatial process has an apparent dimension, visible in the geographical landscape and marked by the heterogeneity of places, but which also shows what can be reproduced. In this case, there is a world of images, forms and appearances that tend towards the homogenisation of our society and which can be better appreciated in the landscape where we see the modes of occupation (CARLOS, 1999).

It can be said that tourist cities "[...] are visual spaces, trapped in the world of images that impose reduction and simulacra. And which reduce appropriation as a commodity of temporary use defined by non-working time" (CARLOS, 1999, p. 176). Still in this theoretical context of analysis, Carlos (1999, p. 178) points out that the space produced increasingly serves the needs of accumulation; therefore, the relations of production "[...] that engender the activities of distribution and consumption are carried out under the aegis of freedom, equality, the reproducible, the repetitive, cancelling out differences in space and time, destroying nature and social time".

In the book Novos caminhos da Geografia, Carlos (1999, p. 178) says that this idea is at the origin of the approach to tourist and leisure urban spaces "[...] produced from reproduction strategies, at a certain moment in the history of capitalism and which increasingly extends to global space", creating new sectors of activity. Rua Rio Preto is not a local tourist urban space, but if the city's public authorities carry out measures to revitalise the built heritage of this place, it is believed that, due to its history, it could be a place of leisure for residents and visitors to the city, where they can find particularities of Quirinopolitan society.

Carlos (1999, p. 179) mentions that tourism represents the conquest of an important part of urban space "[...] that becomes a commodity (and enters the circuit of exchange)", becoming a new profitable branch of productive activity, under this determination. In this sense, urban places "[...] come to have a real existence as a result of their exchangeability through the activity of property developers who use space as a means of reproduction".

The urban spaces that make small, medium and large cities attractive are planned on the basis of:

> A movement that, in turn, is guided by the development of information and marketing, which invents places where people should go on holiday, influencing the consumer with its idea of earthly paradises where everyday life is in apparent suspension (CARLOS, 1999, p. 180).

The changes that occur through the revitalisation of urban spaces constitute a multiplicity of social relationships with spatial practice that underpin the process of generating new commercial opportunities, especially for the old traders.

These spaces maintain their identity with the place and the people who frequent them and escape "[...] the hegemonic homogenising rationality (imposed by the consumer society), which defines a time and a mode of use" (CARLOS, 1999, p. 182). In between:

> [...] public space has a multiplicity of meanings for society depending on culture, habits and customs. The public space can also be a place for communication, dialogue, an extension of living, where children can play,

> young people can date and everyone can expose themselves, talk and make claims. Here, visibility is produced that creates identities - the identity that humanises relationships through the bonds of coexistence and the feeling of belonging to the group and the place (CARLOS, 1999, p. 184).

Furthermore, use as a form of appropriation takes place in the urban space. It is from realities such as Rua Rio Preto, in Quirinópolis, that we find new cut-outs in the geographical discussion of urban space. However, we are not proposing a new construction of space and the new functioning of this territory, based on the destruction of the history that still exists, but rather a debate to think about the preservation and revitalisation of this history.

By avoiding the abandonment by public authorities that was experienced during the fieldwork for this article, the history of this important mobility route in the city is recovered (or, at the very least, preserved). It is essential to continue the historical and current domains of Rua Rio Preto and its surroundings, a territory of neighbouring businesses brought together by a territorial continuity, while the modern settles in other parts of the city. According to Geiger (GEIGER, 1996, p. 238):

> Space itself, such as urban space, corresponds to a higher level of social production. To the extent that the characteristics of this produced space have to do with economic, social, cultural and political structures, they have become the centre of attention for authors (GEIGER, 1996, p. 238).

It can be said that historic and modern urban spaces are appropriated by diverse populations, whether socially or culturally. In cities in the interior of Goiás and Brazil, many neighbourhoods and streets that spring up are named after locals or outsiders who are politically, economically, socially and historically important. Specifically in the case of the section analysed in this article, the name given honours an important drain in the municipality that is a tributary of the Paranaíba River.

Small towns and their commerce: a discussion for thinking about urban Quirinópolis

Urban geography is an area of scientific knowledge that studies cities by defining and theorising them on large, medium and small scales. Quirinópolis can be defined as a small city, as it has approximately 50,000 inhabitants. According to Cavalcanti (2001, p. 12), "[...] geography is one of the sciences that has dedicated itself to analysing the city and urban life. And its categories of analysis have guided a certain way of looking at the city".

With this in mind, this chapter also aims to think about the small town with its social, economic and cultural elements, contributing to other urban perceptions, such as the areas occupied by commercial buildings. Cavalcanti (2001) argues that, in the case of this research, the small town can be understood as a privileged locus of social life, since, more than housing the population, it produces a way of life. Santos (2010, p. 180) points out that in the urban network of the Quirinópolis micro-region:

> [...] the functions or roles of the towns are not so different; most of them provide services to agribusiness, with the exception of São Simão and Lagoa Santa, which also cater for the leisure and tourism sectors. These small urban centres are local towns capable of responding to the real or created vital needs of an entire population, a function that implies a life of relationships.

The economy of these small towns is dynamic and includes agribusiness and the service sector. Santos (2010, p. 180) also states that "[...] the small towns in the interior of Goiás have a significant banking and commercial structure and health professionals. Some have theatres, radio stations and local printed newspapers," as is the case with Quirinópolis. In view of this, the author himself points out another striking feature of commerce in the small towns of Goiás:

> In the small towns in the interior of Goiás, the family groups responsible for small hotels still exist and are the majority, but they have partially sought to make investments focused on specialisation in order to keep up with the complexity and modernisation of the regional hotel sector, signifying a search for professionalisation (SANTOS, 2010, p. 192).

In fact, these characteristics present the commercial changes in these cities. This economic (and, why not, cultural) diversity present in small towns and "[...] the richness of their daily life presents scholars, particularly geographers, with the permanent challenge of understanding them, explaining them and learning about them" (CAVALCANTI, 2001, p. 30). It's worth pointing out that small towns like Quirinópolis also exist:

> [...] with transport facilities and new ways of organising agricultural work, a considerable number of agricultural workers live in the city, which becomes a reservoir of labour. Perhaps this is one of the reasons why, from a certain demographic volume, such locations are able to attract and retain population (SANTOS, 2003, p. 280).

In Quirinópolis, there is still a great deal of consumption in small businesses, based on family or friendship relationships. So do these small towns:

> [...] they provide the means for the final consumption of families and administrations and the intermediate consumption of companies. [...] in reality, with the modernisation of agriculture and the establishment of corporate production, as well as corporate circulation, distribution and information, the demand met by cities ends up being mostly demanded by companies, or in any case this demand is privileged in the establishment of regional plans. [...] the role of the medium-sized cities is to supply the information required by

> agricultural activities immediately and closely, and in this way they become interpreters of the world's technology.

In many cases, this definition is useful for small towns such as Quirinópolis, as urban activity is used to fulfil the close and necessary relationships with regional production, mainly the sugar and alcohol mills.

Quirinópolis brings together "[...] traditional and new activities and professions, as well as harbouring traditional and modern forms of bourgeoisie and middle classes" (SANTOS, 2003, p. 282). Therefore, this small city leads the technical aspects of regional production, leaving the political factors to larger agglomerations in the state of Goiás, such as Goiânia, Rio Verde and Anápolis, due to the role of the metropolis (capital and medium-sized cities) in directly or indirectly driving the so-called global market.

The role of small towns in the political process is limited and incomplete, as well as confusedly perceived. This is not just a reality in medium-sized cities, as Santos (2003, p. 284) states: in smaller and medium-sized cities, "[...] seasonal oscillations can be very clear, reducing employment opportunities. Generally speaking, these seasonal fluctuations are related to agricultural calendars and these to the modern specialisation of production".

In this context, the small urban centres in the interior of Goiás are still places where the patron saint's feast is held, where the rodeos are held, where children play in public squares and where the way of life is very close to rural life. In these cities, there is no concentration of cars (traffic jams), no crowds of people and no need to get up early to get to work, which is located in an urban area far from where you live. There is an everyday life where traditional cultural customs and ways still remain, but without blocking modern development.

Using the words of Corrêa (1997, p. 279), small towns can be characterised as a collection of places with "[...] people, goods, capital and information about geographical space". They are less economically dynamic than medium and large cities. In this sense, it's important to remember that

> Complex patterns of interactions between urban centres have emerged or gained strength, defining a growing interdependence between cities and areas, expressed by interactions: between large cities located in the same region or in different regions; between a large city and smaller centres located both in the hinterland and outside it; and between small cities located in the same region or in different centres (CORRÊA, 1997, p. 283).

It is worth noting that in the geographical networks mentioned by Corrêa (1997) - in other words, the locations articulated with each other by routes and flows, in which spatial interactions effectively take place based on the attributes of the locations and the real possibilities of articulating with each other - there are small towns, but they operate on a different scale to medium-sized towns and metropolises.

The urban commercial centres in this network are different from each other, both as central places and as specialised centres, which predominantly occurs, revealing a strong territorial division of labour between them. Spatial interactions are an integral and traditional part of the geographical theme.

Within this theoretical discussion, shops in small towns emerge, an approach that is also of a geographical nature. In this way, we are referring to the production of the daily lives of people who live in small towns and carry out their activities there, and the spatial arrangements resulting from these activities (leisure, rest, work, education, etc.). With this in mind, it is important to move on to traditional commercial areas from a theoretical analysis that makes it possible to get to Rua Rio Preto.

"On this scale, the analysis can be systematised on the basis of certain elements such as: housing and commercial property, service equipment and socio-professional characteristics" (CAVALCANTI, 2001, p. 13). In the microregion of Quirinópolis, a regional urban reference stands out:

> [...] the marketing of rural products, retail sales, the provision of various services and others are part of the spatial formation that is also organised to serve nearby towns, as a hierarchy is established in the supply of goods and services. This hierarchy is based on the idea of an urban network, defined by a group of nearby towns that establish economic, political and social relations (SANTOS, 2010, p. 179-180).

In contrast to the shopping areas in small urban centres, Lopes (2011, p. 63) mentions that in large and medium-sized cities there are shopping centres and hypermarkets which are generally characterised by the power of attracting consumers to these places, "[...] who are looking for easy parking, mobility, payment, variety, speed and security. These are new forms of commerce that create new centralities, with significant changes to the urban space".

It is notable that commercial areas are responsible for urban dynamics, generating decentralisation, new centralities and a new spatial organisation. Commerce "[...] is responsible for linking the production of goods to their offer for consumption and, consequently, flows" (LOPES, 2011, p. 64). The works analysed indicate that the commercial process is a necessity of urban life. In these terms, the localisation of commerce in general:

> [...] it is discussed by classical theories and current researchers, and is

> intrinsically related to the organisation of urban space. These theories aim to establish theoretical definitions regarding the locational behaviour of economic agents (LOPES, 2011, p. 65-66).

These theories mainly emphasise that, in general, the locational decision for setting up a commercial unit is uncertain, i.e. because the urban environment is unstable and changeable, "[...] betting on a suitable location today does not mean that this same location will have the same advantages in the future" (LOPES, 2011, p. 66). In the specific case of the business on Rua Rio Preto, in Quirinópolis, it is known that its location was associated with the old access to the city, since, in other times, this road was one of the "gateways or access to the city" for those arriving from cities such as Caçu, Cachoeira Alta, among others in the region.

It cannot be denied that commerce is "[...] organised and participates in the reorganisation of urban space through its location and the attraction it exerts" (LOPES, 2011, p. 66). Pintaudi (2008, p. 124) states that "[...] commercial forms are part of the metamorphosis of urban form, which becomes fluid [...]". As such, it can be considered that there are various motivations that lead consumers to certain urban commercial areas, which can be defined as fixed, in the case of Rua Rio Preto.

With technological expansion, "self-service" is common in urban spaces which, in the words of Lopes (2011, p. 69), favours "[...] free access for consumer customers to products, which require standardised packaging". This new commercial model includes supermarkets, hypermarkets and convenience stores. In Quirinópolis, there are the Lucas (Rede), Agrovale and Varejão supermarkets, which are the main ones in the city.

However, none are located in the vicinity of Rua Rio Preto, as these businesses have sought out the most developed, noble and dynamic areas of the city. In Brazil:

> [...] the first records of the emergence of self-service date back to the late 1940s, when Sears was set up in 1949. This shop had self-service characteristics, with modern cash registers, in-store technical assistance and its own car park. Later, in the 1950s, Sirva-se opened in São Paulo, considered to be the first Brazilian supermarket (LOPES, 2011, p. 69).

After the first supermarkets appeared in Brazil, the creation of new units accelerated in all Brazilian regions. Based on the above quote, it can be seen that the emergence of supermarkets in the country is just over six decades old. In these terms, Castro (1997, p. 155) argues that urban commerce can be characterised as part of the social imaginary, visible and interpretable, full of symbolism "[...] present in men's relations with each other and with their environment, which materialise in the different modes of socio-spatial organisation".

Urban commercial property has an undeniable social, economic and cultural value for local people. More than a source of survival for the owners, it is a means of life par excellence, in which the different subjects are unable to leave the spaces full of social relations. This commercial urban space "[...] is both the continent and the content of its signs and symbols" (CASTRO, 1997, p. 157).

There is no vertical urbanisation in the commerce of Rua Rio Preto, but we cannot deny that there is a horizontal extension full of content and density, producing important effects on the individual and collective of the place. These urban commercial relations are present in:

> [...] constructions - streets, squares, monuments, neighbourhoods, blocks, cities - become images, paths and representations of the collective soul. These geographical representations then constitute a way of being, a way of talking about the Earth (CASTRO, 1997, p. 178).

In fact, traditional urban commerce is a fundamental element for understanding urban socio-economic relations, since commercial activity in urban centres goes beyond its classic function of supplying the population with the materials and services it needs: by building a social life between the consumer and the merchant, commerce, as an activity that is clearly inseparable from the urban fact, has always interfered in the ordering of the city, contributing to the shaping of the main urban functional areas.

In this way, commerce has benefited as the town has become more attractive, which is good for both residents and visitors, and brings benefits to the town beyond its economic function and its role as the cultural and social centre of the community. Traditional businesses are almost always located in the city centre and have faced difficulties because of other, more modern businesses. This can lead to a restructuring of relations between traditional establishments in urban centres which, when properly articulated, can take on the characteristics of an open shopping centre.

Rua Rio Preto and its traders

The emergence of the first supermarkets in Brazil, as presented in this chapter, did not lead to the disappearance of urban shopping areas such as those on Rua Rio Preto in Quirinópolis. The establishments that are the subject of this study cannot be included in the new model of commerce highlighted above, as none of them have "self-service" characteristics - the existing car parks are the street itself, i.e. the public space, not the private one.

Many owners have their home attached to their business, and management and service are done on a family basis. With regard to this

Cachinho (1994 *apud* LOPES, 2011, p. 71) states that this model is "[...] centred on the small family business, where the small trader and the traditional shop compete with large sales organisations and new commercial forms". Although the concept of a small town has already been discussed, it is important to highlight the words of Santos (2010, p. 183) - in the small urban centres of the Quirinópolis micro-region, many businesses are:

> [...] simple establishments with family-based management, [...]. Based on ties of consanguinity. In small businesses there are few exogenous workers, and as a result, the notion of inheritance demarcates these places of small private initiatives.

Often, when small businesses come under pressure from the competition of large retail chains, they can't resist the market and are forced to close down. On the other hand, the traditional commercial activity in Rua Rio Preto is still alive and kicking well into the first decades of the 21st century, not as a form of resistance, because one of the things that ensures its existence is the fact that consumers still have the shopping habits of the 20th century.

If consumers remain, investment capital is small and the owners of these shops don't follow the demands of adapting their strategies to new habits, such as using credit cards and personalised bags with the company's brand. Therefore, it can be said that this street is the "[...] place where life develops in all its dimensions, it is the portion of space that can be appropriated for life" (CARLOS, 1996, p. 20-25).

With regard to the location of this commercial area, we can say that it is easily accessible and served by an adequate infrastructure, with power, tarmac and a sewage system. It has a good number of people passing through, but the customer car park and the goods unloading car park are the same. The location makes it easy to cater for a wide audience, belonging to various socio-economic classes, as it is close to the centre.

During the fieldwork, it was possible to understand that the commercial issues of the object of study point to the necessary relationship between urban dwellers and traders. These include the sale and purchase of food, drinks and agricultural equipment, a local tradition, as it is a business that still caters for rural dwellers. As a result of the government's abandonment of the area, it bears negative marks as a place of prostitution and strong violence.

There is, therefore, a complexity of legal and illegal commercial relations in Rua Rio Preto, which has given this article the difficult task of understanding this place, a scientific mission that is not easy given the problems encountered, but not exhausted. There are many visible urban problems, but it is also possible to perceive a place of tradition and history, with its geography full of images recalled by memory and which reconstitute the place researched over the course of the days. Rather than trying to separate the positive from the negative, it was important to take care in the research not to create other prejudices.

Before presenting the results of the fieldwork, with the intention of giving a voice to the traders on Rua Rio Preto, it's worth saying that the trader is defined by Bueno (1996, p. 149) as a "salesman, shopkeeper, dealer". He is the one who has the art of commercialising and serving his clients, who are also called consumers.

The following are the results of a survey carried out among the traders. The survey was carried out between 5 August 2014 and 19 November . The questionnaire was administered to 30 local traders.

Of the business owners, 70 per cent are male and 30 per cent female. There is a predominance of men in the management of businesses, which are small companies or simply family organisations; however, many of them rely on the help of their wives or other family members.

During the fieldwork, some informal testimonies were collected. From the universe surveyed, it emerged that the oldest shopkeeper has been open for 38 years and another interviewee said he had only been here for a month. In addition, 60% of the shopkeepers have had other jobs: they have worked in bicycle workshops, pharmacies, as seamstresses and/or as farm labourers.

Another 40 per cent of those interviewed said that they had always been traders, because most of them learnt the trade from relatives or friends connected to the family and, because they liked the trade, they didn't try to change. According to some respondents, they had even considered changing professions, but due to a lack of opportunities and even resources, these possibilities were discarded.

During the survey, we tried to find out if an old business relationship still exists in Rua Rio Preto: the use of passbooks as a form of commerce. In this type of notebook, the owner of the establishment writes down the names and values of the products purchased by his customers with a set date for payment, which can be weekly, fortnightly or monthly.

Figure 1 shows that 20% of traders sell their products in cash and, surprisingly,

80% still use the traditional passbook. Of those interviewed, none use credit cards, but some confided informally that this change will be necessary because it is a possibility to increase sales.

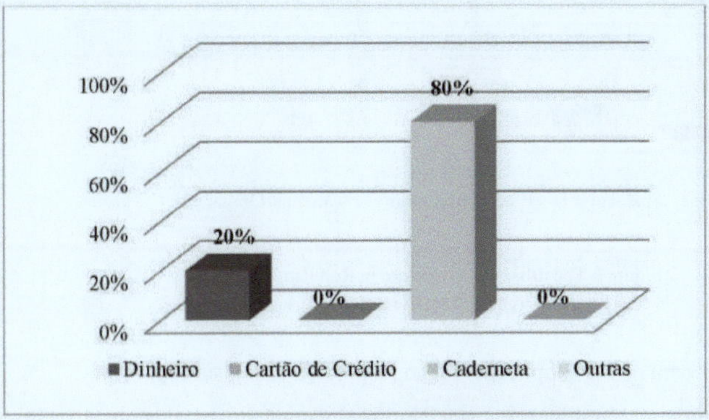

Figure 1 Ways of selling products
Source: Organised by Clemente, 2014.

During the field research, it was possible to see a small number of traders who use credit cards, but we weren't able to talk to them, so the result was 0%, which doesn't fully reflect reality. The habits of modernity are already present in the area, especially the use of mobile phones, the internet and credit cards.

Figure 2 shows the respondents' opinions on the urban infrastructure of Rua Rio Preto. For 80 per cent of those interviewed, improvements are needed mainly in the areas of safety, signposting, lighting and cleanliness. This indicates that the local government has little presence in the , as there is no preservation or revitalisation project on the part of the municipal government.

This urban thoroughfare in Quirinópolis made us think, during our fieldwork, that the city is full of particularities and differences, not just commercial discrepancies, but economic, social and political ones, because Rua Rio Preto seems to have been forgotten by the municipal rulers. If we analyse the city by neighbourhood, we will understand that in the urban area under investigation there are many types of cities, constituting a space of inequalities demarcated by huge differences in interests and forgetfulness. The new always seems to be more important, while the historic becomes an island of abandonment surrounded by the modern.

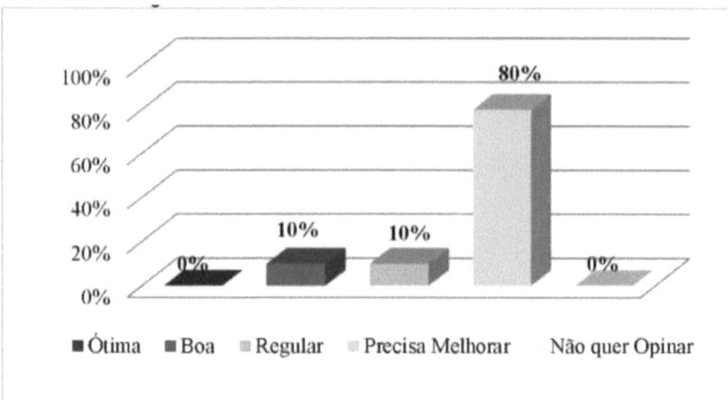

Figure 2 The urban infrastructure in Rua Rio Preto.
Source: Organised by Clemente, 2014.

For 100 per cent of the shopkeepers who responded to the survey, if the residences and commercial buildings on Rua Rio Preto were restored, this place could become attractive to Quirinópolis residents and visitors to the city. According to the data in Figure 3, 80 per cent said that the responsibility for preserving and restoring the local built heritage should be shared between the property owners and the municipal government. However, none of these agents are committed to preserving the town's historical heritage, nor those in the urban and rural areas.

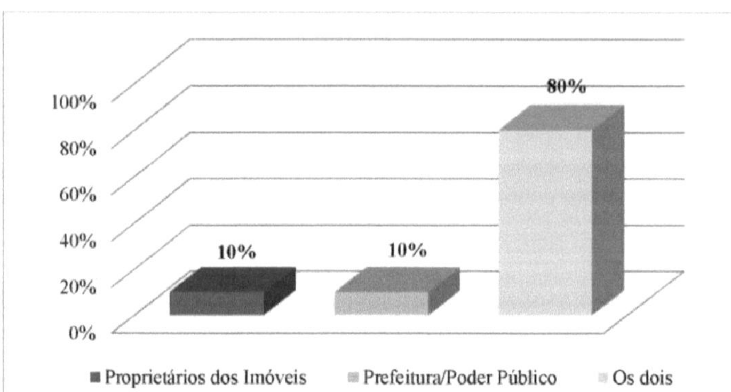

Figure 3 Responsibility for preserving the municipality's historical and built heritage
Source: Organised by Clemente, 2014.

The preservation of historical and cultural assets cannot be the result of just one segment of society, but rather the duty of everyone. The community needs to be aware of the importance of these elements and demand that the local state take action. The concrete actions of social agents are fundamental, as long as they are endowed with interests, strategies and practices that value urban memory. In view of this, 100 per cent of those surveyed were unaware of any municipal law or project to protect the city's built heritage.

During the field research, it was confirmed that, for 100% of the traders, the mobility route investigated is part of the town's history, while, informally, some respondents said that practically all the town's commerce was born on Rua Rio Preto, because this landscape was the main exit route to the countryside and other important towns in the region, such as Caçu, São Simão, Paranaiguara, Jataí and Cachoeira Alta. It was also a departure point for other states, especially Minas Gerais, Mato Grosso and Mato Grosso do Sul - in the past, this street was a commercial centre frequented by many farmers who did their shopping.

According to the field results, 30% of respondents use the city's radio stations to publicise the establishment or, as they put it, "advertise" - this is mainly done on the city's two AM radio stations. Another 40 per cent said that the establishment is known by "word of mouth ", i.e. old and new customers talk to friends, family etc. and this brings in new customers. For 20 per cent, the media work is
through folders and posters, and 10 per cent said they had no publicity work, not even through older, loyal customers.

In the informal reports obtained during the fieldwork, one trader stated that:

> [...] in the 1970s, the mayor of the time forbade the buses that went to the farm to stop at the bus stop there. Then they would stop at the Rio das Pedras bridge, people would do their shopping and the owner of the establishment in Rua Rio Preto would have to walk about 5 kilometres to take his customers' shopping by bicycle to the bus stop. Over time, they began to demand improvements and, with that, the mayor of the time built the bus station here on Rua Rio Preto to improve the conditions for the people who passed through and shopped there. With that came the tarmac, it was very good for us (sic), right? (INFORMAL FIELD DEPOSIT, 2014).

The shopkeepers surveyed said that the best-selling products in their shops on Rua Rio Preto are: clothes, fabrics, medicines, gas, mineral water, coal, drinks, tyres, among others. The bars and the old bus station building are meeting points for country music and, for many interviewees, they are a place for leisure and fun for the locals.

A look at the consumers of Rua Rio Preto

When you think of Rua Rio Preto in Quirinópolis, you immediately think of a traditional shopping centre that aims to supply families living in the urban and rural areas of the municipality. In this vein, Lopes (2011, p. 67) ponders that consumer needs generate "[...] historical changes in the way people trade", but it seems that, in this part of Quirinópolis' urban centre, the customers of the shops on Rua Rio Preto still value the traditions that remain there.

When analysing the object investigated in this article, it is understood that some changes have taken place and have accompanied consumers' new lifestyles, but certain characteristics from the last decades of the 20th century, such as passbook sales, still remain. The results of the field research carried out with 100 consumers who frequent the shops on Rua Rio Preto will therefore be reported here. This methodology was carried out between 5 August 2014 and 19 December of the same year, with the aim of getting to know these people.

The first piece of data collected relates to the gender of the consumers: it should be noted that the majority (60%) are male, mainly because there are many pubs (also known as bars and snack bars) and markets on this urban road, as well as the fact that the place is used as an exit route from the city, making it easier for farmers and other commuters to shop. The main profiles cited by customers of these businesses are: pensioners, housewives, cowboys, salesmen, lorry drivers, farmers and the self-employed.

Regarding the origin of Rua Rio Preto's consumers, the largest number (90%) come from the municipality of Quirinópolis, from the neighbourhoods of Centro, Vila Parreira, Onício Resende and farms, especially from the rural areas of Salgado, Pedra Lisa, Cachoeirinha do Rio Preto and Serra da Confusão. This data shows that the region has a clientele from various parts of the municipality, going beyond the limits of Rua Rio Preto or even its physical agglomeration (the city).

In this context, Figure 4 shows that 53% of consumers have been shopping there for between two and five years. The more traditional customers account for 7%, as they have been frequenting the establishments on this urban thoroughfare for more than a decade - among these individuals, many confided that their grandparents and parents had (and/or still have) the habit of frequenting these shops since the 1970s.

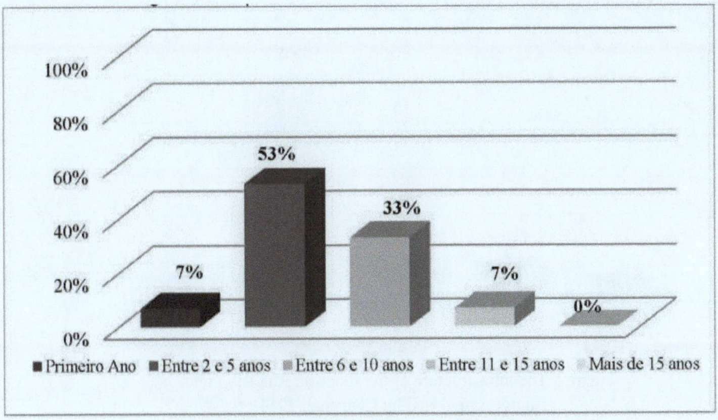

Figure 4 Consumption time on traditional urban roads
Source: Organised by Clemente, 2014.

Meanwhile, in Figure 5, it is possible to understand people's opinions about the infrastructure of the shops on one of the oldest streets in urban Quirinopolis. According to 86 per cent of respondents, the buildings are not in good condition.

and investment in improvements is needed. Many interviewees said informally that the built heritage and the public highway are abandoned and need to be restored and recovered, but they don't believe that any project (or action) will be taken by the owners and local governments.

According to the people who answered the questionnaires, the main products they consume are animal feed, medicines, gas, basic food products, drinks, groceries, beverages, sweets and savoury snacks. The majority (87%) said that they prefer to shop in this area because it is cheaper than other shops in the city. For 93% of consumers, if the urban road and the built heritage were restored and included in public and private cultural agendas, the landscape would attract more people, not just Quirinópolis residents, which would create new forms of use for this place.

None of those interviewed considered that the responsibility for restoration and preservation lies solely with the property owners; however, for 93 per cent, this should be a joint action (society and public authorities), as everyone has to help preserve the city's historical heritage, a part of each resident's life and memory.

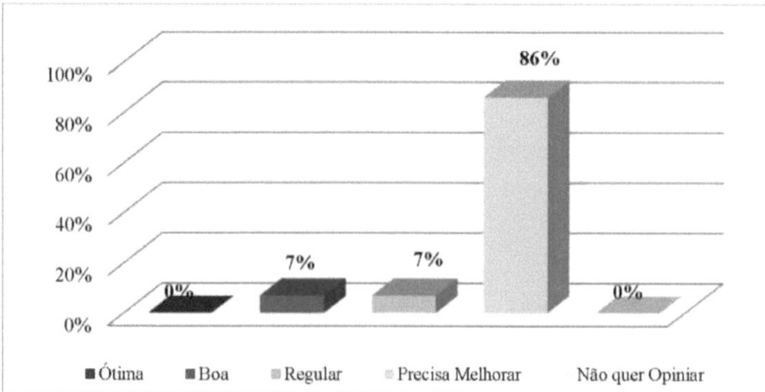

Figure 5 The infrastructure of the shops on Rua Rio Preto
Source: Organised by Clemente, 2014.

Contrary to what is experienced by shopkeepers and consumers on Rua Rio Preto in Quirinópolis, Abreu (2012) argues that after a long period in which only what was new was worshipped - which resulted in a constant and systematic attack on the legacies of ancient times - Brazilian urban daily life is currently invaded by discourses and projects that preach the restoration, preservation or valorisation of various vestiges of the past. In other words, valorisation:

> [...] the past of cities is a common feature of societies **at the turn of the millennium. As far as "new countries" such as** Brazil are concerned, this trend is unprecedented and reflects a significant change in the social values and attitudes that have prevailed until now (ABREU, 2012, p. 19).

For the author, *the* valorisation of the past (or what is left of it in the landscape) is occurring in a generalised way around the world, reflecting the emergence of a new identity relationship between the subjects of the late 20th century and the spatial sets that anchor them on the planet, be they nation states, regions or places. Quirinópolis' public authorities don't look at the past of Rua Rio Preto as one of the most important dimensions of its history, which still lives on in the culture and daily life of the place.

It is essential that public authorities and society in general engage decisively in a movement to preserve what is left of its past, an indication that many aspects have changed in the way society in the interior of Goiás relates to its memories. However, it is not common to find speeches and concrete actions concerned with the past of this city in Goiás.

Built heritage and public neglect

In this work, it is essential to analyse the importance and significance of the built heritage of Rua Rio Preto, as well as the role of managers linked to the public authorities of Quirinópolis. Firstly, Ghirardello and Spisso (2008, p. 13) refer to heritage as "[...] all the material and immaterial, natural or built goods that a person or a people possess or manage to accumulate". Thus, it can be said that architectural heritage is "[...] the buildings that have acquired historical and cultural significance in a given society" (DIAS, 2005, p. 27) - we understand that this is specifically the case with many of the buildings in this article's object of study. The built heritage of Rua Rio Preto is highlighted in the work by Santos (2012, p. 208) on tourism and leisure in Quirinópolis:

> The image shows part of the historic heritage buildings, especially the old bus station on Rua Rio Preto. In this street you can still see houses and businesses built in the mid-20th century, symbolising the urban dynamics of the past.

In his work, the author emphasises that, if well preserved by the local government, this street could become a tourist attraction or a place to visit in the city. In this way, Garófalo and Pinho (2011) point out that public policies, which were considered irrelevant in the past, have now taken on a preponderant and dominant role, as they are fundamental to understanding public space and are linked to human interference, actions and policies of use and appropriation.

From this theoretical point of view, it can be said that the public spaces (squares, streets, leisure areas, public buildings and pavements) of Quirinópolis, especially the built heritage of Rua Rio Preto, are not being preserved, and the public neglect is immense. It is hoped that this work can reach the municipality's various managers and enlighten the public and private sectors, as well as society, about the importance and advantages of revitalising this traditional public thoroughfare in the city, since structuring its commerce and increasing the supply of leisure and cultural spaces can improve the quality of life of its residents.

In fact, revitalisation actions can lead to better urban services and improved urban infrastructure, as well as adding value to commercial buildings and homes that are part of local history. For these actions to take place, the support of municipal leaders and, above all, the community of Quirinopolina is essential.

The private sector involved in the revitalisation project could gain from strengthening cultural, leisure and even tourist activities, which should be included in the revitalisation process. According to Santos (2013, p. 81-82), the public authorities:

> [...] its purpose is to respond to the needs of the resident population in order to

> ensure their well-being. [...] it plays a fundamental role in policies on education, health, security, transport, the environment, tourism, leisure, entertainment, culture and sporting activities, among others. It is the main regulator of spatial use and appropriation, scrutinising and supervising the various economic groups in each new activity created, whose response will determine future decisions in the field of public management and regional development.

For the aforementioned author, it is the role of public authorities, together with society, to plan land use and occupation, accentuate or attenuate regional asymmetries, redefine the structure of cities, define accessibility and other public facilities. These are key issues for an environmental policy that also affect other sectoral policy areas.

According to Santos (2013), people's quality of life is related to their occupation of the territory, through the links between urban and rural spaces, which are often conditioned by incomplete information that has taken place over the years. In these terms, the local government and its organisations are responsible for arranging and revitalising places like Rua Rio Preto.

Given this situation, it can be said that public authorities still play a decisive role in the preservation and conservation of municipal built heritage, injecting public funds. However, it is important to emphasise that:

> However, what is proposed by state policies does not always reach the places, leading subjects to create their own political positions and conceptions of use, and so they start to articulate themselves in different ways, above all based on what is historical and cultural and not according to state political forces, strengthened by the administrative cut-off (SANTOS, 2013, p. 83).

In the specific case of the shops on Rua Rio Preto, it was found that the actions of the local public administration are far removed from the needs of the existing commercial activities carried out by local people, which has strengthened the prejudiced view of a large part of the population of Quirinópolis. There are no projects or targets that are the responsibility of the politicians and officials at the different territorial levels who operate directly in the area, mainly through the administrative bodies of Quirinópolis city hall.

At the local level, commercial competences and whether or not to preserve the built heritage are related to the awareness of each local individual with their territorial roots, but without the technical and financial capacity to create infrastructure and facilities that strengthen commercial relations in the area. It is therefore essential to join forces between the various segments of society to provide knowledge that will lead to the cultural and social development of the area.

Rua Rio Preto needs to be cared for because it is part of the heritage of the city of Quirinópolis (GO), since it was also from this mobility route that the city's commerce was practically born. You could say that it was the beginning of everything, an important part of history that needs to be preserved like any other built heritage. The shopkeepers on this street feel abandoned by the government and are asking for more attention, security and infrastructure.

Final considerations

It is hoped that this research has succeeded in drawing attention to the current situation of areas that have been forgotten by the public authorities in the interior of Goiás, such as the old town centres; the old places where commercial, service and management activities were concentrated; the old industrial areas and residential areas, distinct in terms of form and social content; the old leisure areas; and, among others, those set aside for future expansion. It is essential to think that the urban space of small towns is also transformed through politics and actions in the territory, considered key concepts in the study of Geography.

Specifically in the case of this work, the word "space" was associated indiscriminately with the different scales of the city, the neighbourhood, the street and the house. This space was understood based on relationships that involve cost, time, commerce and historical valorisation. Commercial relations between consumers and sellers are also influenced by whether or not they belong to the same place, neighbourhood or street.

In this discussion, it can be said that small towns organise a role of living, respecting the scales, while the subjects take on a role as traders or consumers in the urban network. In Quirinópolis, Rua Rio Preto, with its products sold in the oldest shops, the old bus station and its built heritage were shown to be an urban asset that is little valued by society and the local government.

Thus, during the fieldwork, it was possible to see that the shopkeepers on this important public thoroughfare feel abandoned by the population and the town hall, mainly because of the lack of security and the fact that the place keeps the old buildings without proper care. In this respect, it was possible to understand that several businesses were closed due to lack of investment or because the owners died and their children didn't want to continue the business, since at the time the crops were more profitable; even so, they didn't sell the buildings, they just rented them out, even though they weren't in good condition. This shows that former local traders were also landowners.

References

ABREU, Maurício. On the memory of cities. In: CARLOS, A. F. A.; SOUZA, M. L.; SPOSITO, M. E. B. (Orgs.). The production of urban space. São Paulo: Contexto, 2012.

BRAGA, R. B. A (re)significificação do conceito de natureza e ambiente no ensino básico: um reflexão sobre as herança iluminista e o trabalho de campo como mediação pedagógica. In: CAVALCANTE, L. de S.; BUENO, M. A.; SOUZA, V. C. de. **The production of knowledge and research into the teaching of Geography**. Goiânia: PUC Goiás, 2011.

BUENO, F. da S. **Dicionário da Língua Portuguesa**. São Paulo: FTD; Lisa, 1996.

CARLOS, A. F. A. **O lugar no/do mundo**. São Paulo: Hucitec, 1996.

CARLOS, A. F. A. The consumption of space. In: CARLOS, A. F. A. (Org.). **Novos caminhos da Geografia**. São Paulo: Contexto, 1999.

CASTRO, I. E. de. Political imaginary and territory: nature, regionalism and representation. In: CASTRO, I. E. de.; GOMES, P. C. da C.; CORRÊA, R. L. (Orgs.). **Geographical explorations**. Rio de Janeiro: Bertrand Brasil, 1997.

CAVALCANTI, Lana de Souza. Geography of the city - the production of urban space in Goiânia. In: CAVALCANTI, Lana de Souza. **Uma geografia da cidade** - elementos da produção do espaço urbano. Goiânia: Alternativa, 2001.

CORRÊA, R. L. Spatial interactions. In: CASTRO, I. E. de.; GOMES, P. C. da C.; CORRÊA, R. L. (Orgs.). **Geographical explorations**. Rio de Janeiro: Bertrand Brasil, 1997.

COSTA, Everaldo Batista. **The concreteness of the tourism phenomenon and commodity heritage cities**: a geographical approach. Rio de Janeiro: Livre Expressão, 2010.

DAMIANI, A. L. Place and the production of everyday life. In: CARLOS, A. F. A. (Org.). **Novos caminhos da Geografia**. São Paulo: Contexto, 1999.

DIAS, Adriana F. **The reuse of built heritage as a protection mechanism**: a proposal for the listed complexes of Florianópolis. 2005. 176 f. Dissertation (Master's in Architecture and Urbanism) - Postgraduate Programme

in Architecture and Urbanism at the Federal University of Santa Catarina, Florianópolis, 2005.

GARÓFALO, G. de L.; PINHO, T. F. de. **Public policies**: limits and possibilities. São Paulo: Atlas, 2011.

GEIGER, P. P. De-territorialisation and spatialisation. In: SANTOS, M.; SOUZA, M. A. A. de.; SILVEIRA, M. L. (Orgs.) **Território** - globalização e fragmentação. São Paulo: Hucitec, 1996.

GHIRARDELLO, N.; SPISSO, B. **Historical heritage**: how and why to preserve it. Bauru: Canal Seis, 2008.

LIMONAD, Ester. Urbanisation and the organisation of space in the age of flows. In: SANTOS, Milton; BECKER, Bertha K. **Territórios, territórios** - ensaios sobre o ordenamento territorial. Rio de Janeiro: Lamparina, 2007.

LOPES, Michelly de Lourdes. **The spatialisation of retail in intra-urban space**: the case of the Bretas supermarket chain in Uberlândia (MG). In: ALVES, L. A.; RIBEIRO FILHO, V. O espaço intraurbano de Uberlândia (MG): perspectivas geográficas. Uberlândia: Edibrás, 2011.

MOURA, Poliana S.; SILVA, Mirella L. Fieldwork in the tourist landscapes of the destination Canastra - Minas Gerais. In: SANTOS, Jean, C. V. (Org.). **Landscapes and tourist destinations in geographical research**. Uberlândia: Composer, 2009.

OLIVEIRA, Francielle F. de; BUENO, Kamilla F. Student tourists in the city of Lagoa Santa (Goiás): from the thermal waters to the field report. In: SANTOS, Jean, C. V. (Org.). **Landscapes and tourist destinations in geographical research**. Uberlândia: Composer, 2009.

PINTAUDI, S. M. The consumption of consumer space. In: OLIVEIRA, M. P.; COELHO, M. C. N.; CORREA, A. M. (Orgs.). **Brazil, Latin America and the world**: contemporary spatialities. Rio de Janeiro: Lamparina; FAPERJ; ANPEGE, 2008.

SANTOS, Jean Carlos Vieira. **Regionalisation policies and the creation of tourist destinations between Lago de São Simão and Lagoa Santa in the Lower Paranaíba region of Goiás**. 367 f. Thesis (PhD in Geography) - Postgraduate Programme in Geography, Institute of Geography, Federal University of Uberlândia, Uberlândia, 2010.

SANTOS, Jean Carlos Vieira. Tourism, leisure and attractions: contemplations, uses and appropriations of landscapes. In: URZEDO, M. da F. A. **Quirinópolis** - mãos e olhares II: história & imagem. Goiânia (GO): Kelps, 2012.

SANTOS, Jean Carlos Vieira. **Region and tourist destination**: sensitised subjects in the geography of places. São Paulo: All Print, 2013.

SANTOS, Milton. **O Brasil:** território e sociedade no início do século XXL Rio de Janeiro: Record, 2003.

CHAPTER 7

SPATIAL REORGANISATIONS IN THE MUNICIPALITY OF QUIRINÓPOLIS/GO AND FEMALE PROSTITUTION IN 2013[47]

LEÃO, Lucas [48]
SILVA, Lorranne Gomes da[49]

Introduction

The arrival of agribusiness, especially in 2008 in the municipality of Quirinópolis, was the key to the spatial reorganisation of the city and the countryside. It can be said that the municipality has been and is being "captured" by national and international capital, which, by investing heavily, has caused drastic changes to local landscapes and society.

The aim of the research was to reflect on the spatial reorganisation of the municipality of Quirinópolis and female prostitution. The initial motivation for carrying out research in this area came from the possibility of providing clarification to Quirinopolis society and showing aspects of this profession that is so old and ignored, even by the public authorities.

The geographical basis of the work was anchored in the category of **space** analysis, thus referring to spatialisation. The aim was to understand whether all the development that the city has undergone has also led to an increase in female prostitution.

The map of prostitution spaces was enlightening for the reflections. The interviews and testimonies were fundamental to realising the hypotheses and authors such as Santos (1997, 1998), Júnior (2012), Arruda (2013), among others, were fundamental to the writing.

1 The city of Quirinópolis: spatial reorganisations

Space is the central element for the production of human life. In it, societies and

---

[47] This article is the product of monographic research defended and approved at the State University of Goiás/Campus of Quirinópolis in November 2015, as part of the Geography course.

[48] Graduated in Geography at the State University of Goiás, Quirinópolis Campus. .lucasleao21106@hotmail.com

[49] Professor of Geography at the State University of Goiás. PhD in Geography from the Federal University of Goiás, Institute of Socio-Environmental Studies, Goiânia Campus. .lorrannegomes@gmail.com

their different peoples organise themselves according to their customs, culture and way of life. In this way, spatial understanding was fundamental to the analyses of female prostitution in the city of Quirinópolis/GO. Thus, it was through space that the analyses were constituted.

We live in the era of globalisation, of accelerated scientific and technological growth, of the universalisation of society, of virtual capital, of a demand for consumer goods that is proportional to population growth and the accumulation of wealth and material goods. According to Santos (1997, p.15):

> [...] a form of universalisation that alters the geographical dimensions of human activity, turning the world into a global capital space, aggravating technological and organisational disparities between places and accelerating the process of economic and geographical concentration.

In this context, space becomes a universal commodity and makes it possible for different places to be globalised by capital, a fact that can be seen in countless Brazilian municipalities. The municipality of Quirinópolis, the locus of the research, could be analysed through the dynamics of capitalist production - local/global. Quirinópolis is a Brazilian municipality located in the south-west of the state of Goiás, belonging to the mesoregion of southern Goiás, as shown in map 01:

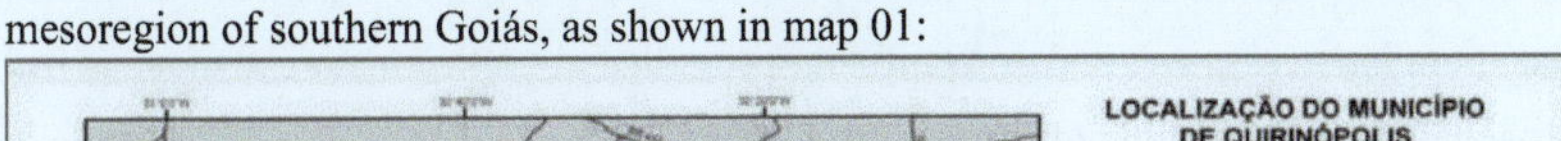

Map 01: location of the municipality of Quirinópolis/GO.
Source: SIG-Goiás. Prepared by: Vinícius G. de Aguiar, 2010.

The municipality covers an area of 3,780 km^2. Its area represents 1.1115 % of the state's area. Of this total, 84.42 % of its population live in urban areas, while 15.58 %

live in the countryside.

Quirinópolis has undergone spatial reorganisations, especially since 1998, as a result of the implementation of agribusiness activities. The replacement of traditional peasant agriculture by agribusiness activities has changed the productive dynamics of the municipality, as well as the population, both in the countryside and in the city. According to Padilha Junior (2012, p.11):

> Agribusiness is the sum total of production and distribution operations for agricultural supplies, production operations at agricultural units, storage, processing and distribution of agricultural products and items produced from them.

Agribusiness involves the production, industrialisation and marketing of agricultural supplies, i.e. all scales of the production sector, including marketing, service provision, etc. According to Fernandes and Welch (2008), agribusiness is the new name for the economic development model of capitalist agriculture. In the 21st century, it is one of the ways in which capital manifests its power of (re)production by conquering geographical space.

Agribusiness has established itself in several Brazilian municipalities, and ends up (re)producing new social relations of production by creating new productive sectors, incorporating new spaces and/or reunifying spaces that at another historical moment had separated them. According to Arruda (2013, p.74):

> In all the sectors that make up agricultural production activities, there is an increasing rationalisation of production processes, which is permeated by the social and technical organisation of work in order to accelerate productivity and increase the conditions for producing surpluses and earning profits. This production involves different capitals and labour processes in different places, but they are concomitantly linked to the market through the generation of value.

Thus, with the globalisation of capital, manifested in activities linked to agribusiness, the volume and intensity of transactions between cities has increased, making them strategic locations for the global economy, playing a supporting role in these activities and making the relationship between the local and the global more inseparable, as is the case with Quirinópolis.

The establishment of the mills in the municipality of Quirinópolis can be explained not only by natural reasons, which attracted the attention of commercial groups from the south-east of the country, but also by the logistical issue in which the municipality is located, making it a production outlet for various parts of the country, according to Santos (2015, p. 6):

> Logistics also rely on the road network: BR 153 and 452 and also GO 164, which connects to BR 452 (Paranaiguara) and GO 206, which connects to BR 384 (Itumbiara, another large sugar-alcohol complex), as well as the proximity of the alcohol pipeline that heads eastwards to Senador Canedo (GO), towards the metropolitan region of Goiânia.

The establishment of the São Francisco and Nova Fronteira Bioenergia sugar and ethanol plants has brought reorganisation to the city at all levels. After the plants were set up, Quirinópolis began to maintain local, regional and national relations, as well as with distant centres (the United States, Japan, Sweden) through the transactions established by the agro-industrial corporations present in its space and through an urban economy based on the export of agricultural **commodities**, which, according to Branco (2008, p.12):

> is the term used to refer to products of primary origin that are traded on commodity exchanges. They are usually products in their raw state or with a small degree of industrialisation, with almost uniform quality and are produced and marketed in large quantities from a global point of view. They can also be stored without significant loss of quality over a certain period. They can be agricultural, mineral or even financial products.

Considering this new economic dynamic in the municipality, the demographic growth is significant. According to data from the IBGE (2010), in 2000 Quirinópolis had 36,512 inhabitants, in 2010 the population jumped to 43,220 and in 2016 there were an estimated 47,950 inhabitants.

The arrival of agribusiness reshaped several local factors, which added up to positive and negative points for the town. Among the positives, we can highlight: an increase in population (the municipality became attractive to labour, attracting migrants from other regions, especially the north-east); an increase in the number of jobs; an increase in commercial activities; a boom in the real estate sector; a better quality of life, due to the increase in residents' per capita income; an increase in consumption, attracting the arrival of new companies; among others.

Among the negatives are: an increase in violence; theft and prostitution; an increase in the cost of living (it must be considered that the ratio between the growth of the municipality and people's income is disparate and unequal, since not all of them are included in the new productive dynamic, which leads to an increase in social inequality); an increase in air and water pollution; the weakening of the soil; the "expulsion" of peasants from the countryside; the disorganised growth of the city; among others.

Given the brief context presented, it can be seen that while, economically, the municipality is of great importance to the state's economy and the new dynamic has improved the quality of life for many Quirinopolinos, on the other hand, the social ills suffered by many people are evident. Many are excluded from the process, in addition to the environmental consequences that have yet to be calculated by the "owners of capital". The following subtopic deals with female prostitution in this context.

1.2 - The spatialisation of female prostitution in Quirinópolis/GO

Human spatial production generates new concepts and characterisations that can be analysed, including spatiality and spatialisation. According to Santos (1998), spatiality is a moment of geographised social relations, the moment of society's impact on a given spatial arrangement.

Spatialisation is the functional and structural forms of spaces, which is why it can be mutable, circumstantial and subject to change. It shows the present, but it is not the result of society's movement alone, because it depends on space in order to be realised, since in its permanent movement, in its incessant search for geography, society is subordinated to the law of pre-existing space (Santos, 1998).

In this way, we sought to spatialise female prostitution in the city of Quirinópolis. Through interviews and informal conversations with residents of the city, we found that there has been an increase in houses or spaces of prostitution,

especially after 2008, when the sugar and alcohol mills arrived, which may have contributed to the increase in female prostitution.

With the arrival of mostly male employees, places specialising in sexual activities multiplied, thus portraying the old law of "supply and demand". Interviewee A said

> The city has grown a lot, and many people from outside have come to live here. In the past, there weren't many cars in the city, but today the reality is different: there are lots of cars, as well as a great demand for rented accommodation, and commerce has also grown. I think that with so many men from outside the city, the demand for these women is much greater.

Interviewee B said:

> The mill injected a lot of money into the town, which attracted people from other towns, most of them to work in the mills. Other companies also came to provide services, as well as pharmacies, supermarkets and also brothels, as the town received many men.

According to the data produced by the surveys carried out in the research, in mid-1998, when the sugar-alcohol plant developments were being set up, it is estimated that there were approximately five brothels recognised by society. In 2013, twelve prostitution centres were mapped around the urban perimeter, as shown in map 02:

Reading the map shows that prostitution spaces are diverse and can be differentiated as points of prostitution, bars and brothels.

Most of the time, bars are just a meeting place where prostitutes have contact with their future clients. Prostitution points are characterised by street spaces, such as street corners or in front of busy places. Brothels are specialised houses of prostitution. In them, girls are available to receive clients; there are bars inside, as well as rooms where the girls have intimate encounters with their clients.

In short, the purpose of prostitution centres is to attract customers, and that is why most of them are located in strategic and easily accessible areas. Of the twelve spaces catalogued, twelve were identified as points of commerce. This data shows that, due to the dynamism of nightlife, the number of places where there is some kind of prostitution tends to be higher than the number surveyed, since there may be spaces that are evidently not considered for this particular purpose - prostitution.

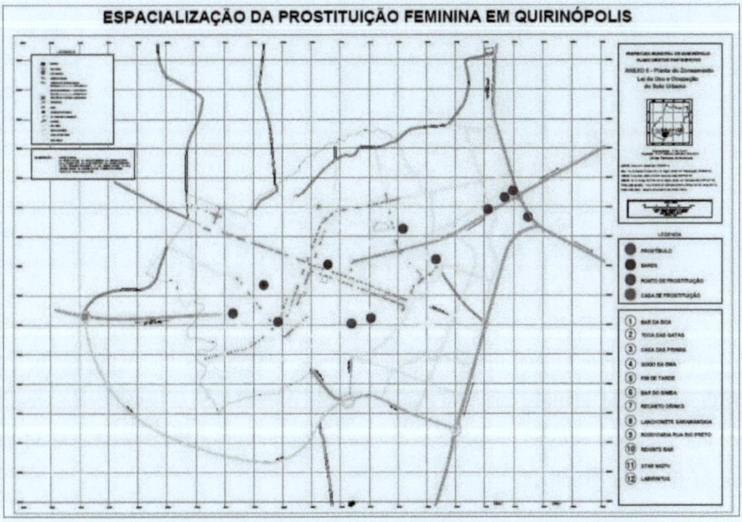

Map 02: The spatialisation of prostitution in Quirinópolis/GO.
Source: Quirinópolis City Hall

Another factor that alters the data presented is clandestine prostitution centres, to which few people have access and, in some cases, are not even aware of their existence. Unfortunately, the sexual exploitation of children and under-18s is common in clandestine spaces, since there is no type of inspection or penalties for those who commit such offences.

After investigating the female prostitution centres, an attempt was made to draw up a profile of these women, which will be presented below.

1.3 - The profile of female prostitution in the city of Quirinópolis/GO in 2013

During the research, the need arose to get to know or draw up a profile of the call girls who work in Quirinópolis. To this end, interviews were recorded in order to gather information about the prostitutes' profession and work. The aim was not to generalise about these women, but to find out a little more about the reality in which they live.

Unfortunately, due to the marginalisation of this profession, many refused to answer the questions and preferred not to take part in the survey. It was possible to interview only 16 women[50] , aged between 18 and 32, as shown in Graph 01:

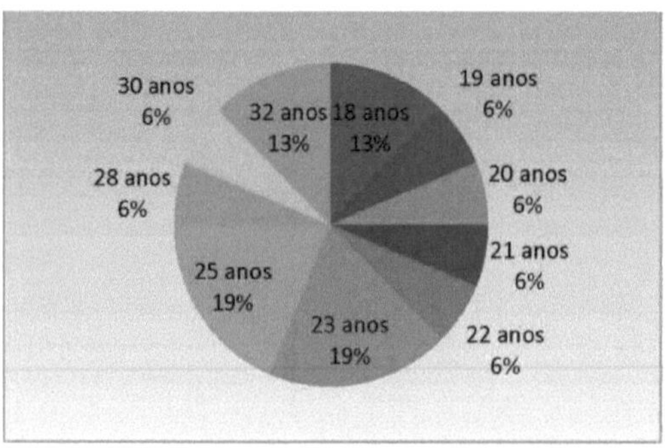

Graph 01: Age of the interviewees
Source: LEÃO, Lucas Clara , Fieldwork, 2013

It is therefore known that it is quite common to find girls under the age of 18 among them. In fact, another question that was asked proved this. When they answered at what age they had entered prostitution, it was found that the majority had entered as minors, as shown in graph 02:

[50] Less than half of the interviews proposed at the beginning of the research.

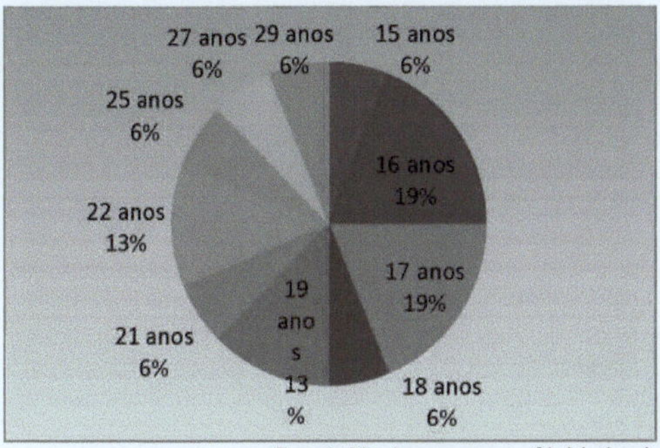

Graph 02: Average age of initiation into prostitution
Source: LEÃO, Lucas Clara ,Fieldwork, 2013

The highest results were obtained for ages 15, 16 and 17, which together account for 44 per cent of the total. Of these, 32 per cent said they had started prostituting between the ages of 19 and 22. Only 18 per cent said they had started in the profession between the ages of 25 and 29.

Most of the women interviewed had studied very little and the reasons for this were varied. It was noted that 44 per cent had not completed primary school, 12 per cent had completed primary school, 31 per cent had completed secondary school and 13 per cent had completed higher education.

Among the questions in the questionnaire, the women's place of birth was important in revealing the existence of a prostitution network. The answers show that most of the women who work as prostitutes in the city of Quirinópolis come from other states (Minas Gerais and São Paulo) and/or other cities in the state of Goiás (Rio Verde, Jataí, Santa Helena, Goiânia).

If it's difficult to control local prostitution, when it turns into a regional, national or even international network, it's even more difficult, because it's known that it's through networks that other problems, such as child prostitution, trafficking in women, sexual slavery, among others, begin to appear.

According to interviewee C, the most frequent reasons why these women leave their hometowns to prostitute in Quirinópolis are: "to look for different places to work, because over time we stay in the same place for a long time and end up not attracting any more clients, so I always change locations and cities". For interviewee D: "Quirinópolis, because it has industries like the power station, attracts a lot of male workers, and because of the money the city has. It's a rich city, and we come in search of money, the more the city has, the better".

Money is the great attraction of the profession and the goal of the girls who prostitute themselves. The survey found that the average monthly earnings of prostitutes in Quirinópolis is between R$2,000 (two thousand reais) and R$2,800 (two thousand eight hundred reais), as shown in graph 03:

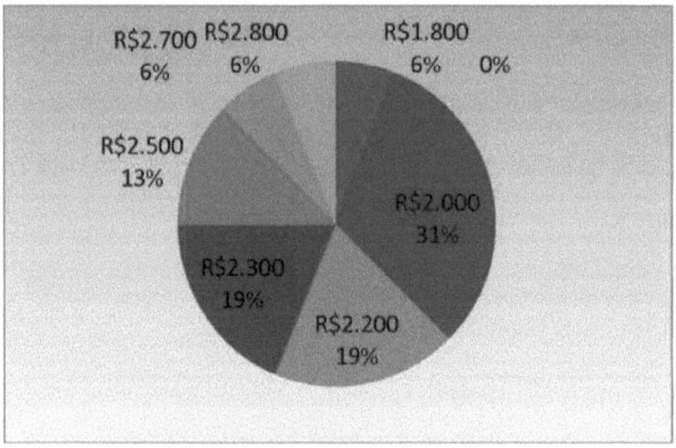

Graph 03: Monthly earnings of prostitutes
Source: LEÃO, Lucas Clara, Fieldwork, 2013

The fact that the girl has to share her earnings with brothel owners means that the amount charged for the programme doesn't go entirely to her. According to the interviewees, the economic factor is the most common determinant of entry into prostitution, followed by the end of marriage and abandonment by the family, associated with the difficulty of integrating into the labour market.

For interviewee E: "It's very difficult when you find yourself alone, without a family. My separation from my husband left me with no direction, so I opted for this life". Interviewee F said: "with no studies and no work experience, my only alternative was to prostitute myself, because I couldn't find work and I needed food and a place to live".

There is often an expectation on the part of women that their stay in prostitution will be temporary, fuelled by the hope of getting another job, going back to school, finding a partner and getting married. Thus, for many sex workers, prostitution is still seen as a short-term strategy, coinciding with the transience of the difficulties they face in maintaining themselves and their children.

However, this rarely happens, as most of them know when they started, but have no idea when they will leave and end up being dependent on the profession for a long time, even though they don't like what they do. Of the women interviewed, 69 per cent were categorical in answering that they are not satisfied with the life they lead and that they intend to leave the profession one day. On the other hand, 31% answered that they feel good and satisfied with their work - most of them work in brothels (75%) - and some of them do prostitution in street spots (19%) and bars (6%).

From what was presented in the data, in the interviews and in the recorded testimonies that were not disclosed in the research, at the request of the interviewees, it can be concluded that prostitution in Quirinópolis has its own profile which, consequently, is related to the local reality. The city has grown a lot in the last decade, especially after the agribusiness ventures, the demand for jobs, and the migration of people, especially for temporary work in the mills, which has increased significantly.

It can be said that the migrant men interviewed, who leave their states and/or cities to work in the sugar and alcohol mills in Quirinópolis, are not rooted in the place, i.e. their stay is sporadic and usually lasts between six months and a year. These men usually look for prostitutes, as they only want to satisfy their momentary sexual desires. This is one of the reasons for the increase in prostitution centres and prostitutes in Quirinópolis.

Most prostitution spaces are not obvious in the city; they use bar, warehouse, bus station and bus stop façades to camouflage their true functions. This shows that identifying these points can be more difficult than it seems, requiring a longer research period.

The average salary of prostitutes is significant, and for this it is important to emphasise the local growth, as already mentioned, as well as the increase in the **per capita** income of Quirinopolinos. The interviewees also emphasised that they chose Quirinópolis precisely because of the possibilities of earning more money.

Another new factor exposed in the data was the development of a prostitution network linked to cities in the state of Goiás and other Brazilian states. It is known that this network tends to strengthen and promote the arrival of other call girls in the future.

Although most of the women interviewed say they don't like what they do, most of them have been in the profession for more than ten years. This contradiction raises many questions about what really holds these women to this profession that is so repudiated by society, where they are exposed to illness, mental problems, violence and so many other situations.

The government's disregard for prostitution is notorious. There are few initiatives for the social inclusion of these women in society, and there is minimal monitoring of the risks suffered by their "commodity" bodies - health, abortion, rape or any other cause of violence. In Quirinópolis, there are no public policies aimed at improving the lives of these women.

In fact, the same 15th century prejudice persists in the lives of these women, regardless of the society in which they live. Most people condemn and even attempt acts of violence against them. Although there is a law in Brazil that legalises this profession, it does not guarantee in practice the rights that these individuals have to be dignified in the work choices they have made for their lives. Being a prostitute - in addition to all the difficulties that surround it - can be a choice made by each individual.

If they are choices, we need to accept and respect what each person wants for their life, and if they are conditions of marginalisation, survival or need, governments, laws and public policies need to act in favour of these individuals, promoting their rights and their inclusion as citizens.

Therefore, in addition to trying to build up a spatialisation of prostitution in Quirinópolis and the profile of prostitutes, the main aim of this research is to be a voice in the fight against the marginalisation of these women and any kind of prejudice.

Final considerations

The research sought to highlight the spaces created by prostitution in the city of Quirinópolis, representing, through a map, how this activity is spatialised. It also looked at the profile of the call girls who offer their services in the city, through interviews and testimonies.

It can be seen that the number of brothels in the city has increased significantly, spreading to different parts of the city. This may be linked to the fact that the municipality is home to agro-industrial centres, which, as in the case of shops, has increased the number of brothels and attracted sex workers from a wide variety of sectors. The neighbourhoods where prostitution is present are mostly poor. In the city's upmarket neighbourhoods, no traces of prostitution were found, either as brothels or on the street itself.

Most of the call girls have only completed high school and are aged between 18 and 30; many said they are not satisfied with their work and want to leave prostitution, their current condition, in the near future.

Quirinópolis is currently linked to a prostitution network, and this is mainly due to the number of men that the city concentrates for agribusiness activities, attracting call girls. The impact on the local economy is not so clear-cut, but it is relevant when you analyse the earnings figures, which are mostly above two thousand reais per girl.

In addition to drinks and renting rooms in brothels, it's clear that the market generates a lot of money, although we can't give exact figures. Although it's hardly noticed, prostitution in the city of Quirinópolis generates a lot of capital and is an important factor in the local economy.

The work has therefore been able to show how the economic factor crystallised by capital has influenced the city's sex market and how it has been spatialised. In this way, it can be concluded that the mediator of the new situations that the city has come to receive and visualise is the result of the high turnover of money, which has led to an increase in the quality of life of its population. As a result, markets such as the sex industry took advantage of the opportunity to gain a foothold and thus grow.

References

ARRUDA, Zuleika Alves de. **Cities in the region of agribusiness expansion and new territorialities.** Territorium Terram Magazine. V.01, N° 01, p. 72-94 | Oct/Mar - 2012/2013

BRANCO, André Luis de Oliveira Castello. **Soya production in Brazil:** an econometric analysis for the period 1994-2008. 2008. 54 f. Course Conclusion Paper (Graduation) - Faculty of Economic Sciences, Pontifical Catholic University of Campinas, Campinas, 2008.

CARTLEDGE, Paul. **Democritus: Democritus and the Politics of Atomism**. [translated by Angélica Elisabeth Kohnke]. São Paulo: Editora UNESP, 2000.

DOLLFUS, Olivier. **Geographical Space.** 4ª Ed. Difusão editorial: São Paulo, 1982.

IBGE, **Brazilian Institute of Geography and Statistics** (2012).

FERNANDES, B. M and WELCH, C. A. The orange agribusiness in Latin America. In: FERNANDES, B. M. (org.) **Campesinato** e agronegócio **na América Latina:** a questão agrária actual. 1st d. São Paulo: Expressão Popular, 2008.

LEAL, Maria Lúcia; LEAL, Maria de Fátima P. (eds). **Research on trafficking in women, children and adolescents for the purposes of commercial sexual exploitation - PESTRAF: national report** - Brazil. Brasília: CECRIA, 2002.

LIBANIO, João Batista. **The Logics of the City.** 2nd ed. São Paulo: Edições Loyola, 2001.

MORAES, Antonio Carlos Robert. COSTA, Wanderley Messias da. **The valorisation of space.** 4th ed. São Paulo: Editora Hucitec, 1999.

PADILHA JÚNIOR, J. B. **Agronegócios**. Federal University of Paraná, Agricultural Sciences Sector, Department of Rural Economy and Extension -DERE. Paraná: UFPR, 2012. Available at:https://pt.scribd.com. Accessed: 20/09/2016.

ROSSIAUD, Jacques. **Prostitution in the Middle Ages**. Rio de Janeiro: Paz e Terra, 1991.

ROUSSELE, Aline. Poneia: **Sexuality and love in the ancient world**. São Paulo Brasiliense, 1983.

SANTOS, Milton. **Thinking the Space of Man.** 3rd ed. São Paulo: Editora Hucitec, 1997.

_______________ . **METAMORPHOSES OF THE HABITED SPACE: theoretical and methodological foundations of geography**. Hucitec.São Paulo 1988.

RECKLESS, W. C. Distribution of organised prostitution throughout the city. In: PIERSON, D. (Org.) Estudos de ecologia humana. São Paulo: Martins Editora S. A., 1976. p. 480-498

SANTOS, Milton. **Space and Method.** 1st ed. São Paulo: Editora Nobel, 1985.

________. **For a new Geography.** 6th ed. São Paulo: Editora Edusp, 2002.

SANTOS, Vonedirce Maria. The foundations for the consolidation of a new economic centrality: ethanol from the Cerrado and the agro-energy complex in the Quirinópolis micro-region. In: JESUS, Jose Novais de; SANTOS, Gilberto Celestino dos (eds.) **Geography and Subjects of the Cerrado:** analyses and reflections, Goiânia: Kelps, 2015.

MIX
Papier aus verantwortungsvollen Quellen
Paper from responsible sources
FSC® C105338

Printed by Books on Demand GmbH, Norderstedt / Germany